Energy, Environment, and Sustainability

Series Editor

Avinash Kumar Agarwal, Department of Mechanical Engineering, Indian Institute of Technology Kanpur, Kanpur, India

AIMS AND SCOPE

This books series publishes cutting edge monographs and professional books focused on all aspects of energy and environmental sustainability, especially as it relates to energy concerns. The Series is published in partnership with the International Society for Energy, Environment, and Sustainability. The books in these series are edited or authored by top researchers and professional across the globe. The series aims at publishing state-of-the-art research and development in areas including, but not limited to:

- Renewable Energy
- Alternative Fuels
- Engines and Locomotives
- Combustion and Propulsion
- Fossil Fuels
- Carbon Capture
- Control and Automation for Energy
- Environmental Pollution
- Waste Management
- Transportation Sustainability

Review Process

The proposal for each volume is reviewed by the main editor and/or the advisory board. The chapters in each volume are individually reviewed single blind by expert reviewers (at least four reviews per chapter) and the main editor.

Ethics Statement for this series can be found in the Springer standard guidelines here https://www.springer.com/us/authors-editors/journal-author/journal-author-helpdesk/before-you-start/before-you-start/1330#c14214.

Nikhil Sharma · Camille Hespel ·
Snehasish Panigrahy · Paramvir Singh ·
Avinash Kumar Agarwal
Editors

Sustainable, Scalable and Storable E-Fuels for Decarbonising Transport Sector

Springer

Editors
Nikhil Sharma
Department of Mechanical Engineering
Malaviya National Institute of Technology
Jaipur
Jaipur, Rajasthan, India

Snehasish Panigrahy
Department of Energy Science
and Engineering
Indian Institute of Technology Delhi
New Delhi, Delhi, India

Avinash Kumar Agarwal
Department of Mechanical Engineering
Indian Institute of Technology Kanpur
Kanpur, Uttar Pradesh, India

Camille Hespel
PRISME Laboratory
Université d'Orléans
Orléans, France

Paramvir Singh
Department of Mechanical Engineering
National Institute of Technology Agartala
Agartala, Tripura, India

ISSN 2522-8366 ISSN 2522-8374 (electronic)
Energy, Environment, and Sustainability
ISBN 978-981-96-7387-2 ISBN 978-981-96-7388-9 (eBook)
https://doi.org/10.1007/978-981-96-7388-9

This Springer imprint is published by the registered company Springer Nature Singapore Pte Ltd.
The registered company address is: 152 Beach Road, #21-01/04 Gateway East, Singapore 189721, Singapore

Preface

E-fuels are synthetic fuels made from renewable electricity and captured carbon, offering a sustainable solution for decarbonizing transport. They can be scaled using existing infrastructure and stored for long periods, making them versatile for use in sectors like heavy-duty transport, aviation and shipping. Despite challenges such as high production costs, e-fuels help reduce emissions in various sectors. They also enhance energy security, create jobs and reduce reliance on imported oil. E-fuels offer a cost-effective way to transition to cleaner energy while supporting a sustainable, low-carbon future.

The International Society for Energy, Environment and Sustainability (ISEES) was founded at the Indian Institute of Technology Kanpur (IIT Kanpur), India, in January 2014 to spread knowledge and awareness and to catalyse research activities in the fields of energy, environment, sustainability and combustion. Society's goal is to contribute to the development of clean, affordable and secure energy resources and a sustainable environment for society, spread knowledge in the areas mentioned above and create awareness about the environmental challenges the world is facing today. The unique way adopted by ISEES was to break the conventional silos of specialisations (engineering, science, environment, agriculture, biotechnology, materials, fuels, etc.) to tackle the problems related to energy, environment and sustainability holistically. This is evident by the participation of experts from all fields to resolve these issues. The ISEES is involved in various activities, such as conducting workshops, seminars, conferences, in the domains of its interests. The society also recognises the outstanding works of young scientists, professionals and engineers for their contributions by conferring awards in various categories.

The Eighth International Conference on "Sustainable Energy and Environmental Challenges" (VIII-SEEC) was organised under the auspices of ISEES from 04 to 06 December 2023 at the Malaviya National Institute of Technology Jaipur, India. This conference provided a platform for discussions between eminent scientists and engineers from several countries, including India, USA, Brazil, Canada, France, Greece and Africa. At this conference, eminent international speakers presented their views on energy, combustion, emissions and alternative energy resources for sustainable development and a cleaner environment. The conference presented two high-voltage plenary talks by Prof. G. D. Yadav, Institute of Chemical Technology, Mumbai, and Prof. Norma A. Alcantar, University of South Florida, USA.

The conference included 21 technical sessions on energy and environmental sustainability topics, including two plenary talks, 23 keynote talks, 35 invited talks from prominent scientists and 95 contributed talks by students and researchers. The conference included technical sessions on alternative fuels for sustainable transport, advanced engines and fuels, sustainable technologies, new energy, biomass and gasification, advanced diagnostics and optical techniques, policies for environmental sustainability, solid waste remediation, wastewater treatment, air pollution and control, natural treatment system, pollution mitigation, environmental sustainability: challenges and priorities, energy and environment, sustainable processing of biomass, biofuels and biorefineries, environmental bioengineering, microbial processes, bioremediation, human health and environmental sustainability and several assorted topics.

250+ participants and speakers attended the conference, where six ISEES books published by Springer, Singapore, under a dedicated series, *Energy, Environment and Sustainability*, were released. This conference laid the roadmap for technology development, opportunities and challenges in the energy, environment and sustainability domain. These topics are relevant for the country and the world in the present context. We acknowledge the support from various funding agencies and organisations for the successful conduct of the VIII-SEEC, where the idea of these books germinated. We would therefore gratefully like to acknowledge Malaviya National Institute of Technology Jaipur (special thanks to Dr. Nikhil Sharma); SERB, Government of India; Department of Scientific and Industrial Research (DSIR) (special thanks to Dr. Vipin Shukla), Sponsors (Dynomerk Controls India Pvt. Ltd., Rebound Envirotech Pvt. Ltd., BlueDrop Enviro Pvt. Ltd.) and our publishing partner Springer (special thanks to Ms. Swati Mehershi).

The editors would like to express their sincere gratitude to many authors worldwide for submitting their high-quality work on time and revising it appropriately at short notice. We want to express our special gratitude to our prolific reviewers: Dr. Sujeet Yadav, Prof. Dilip Sharma, Dr. Nidhi Sharma, Dr. Anoop Kumar Shukla, Dr. Rohit Kumar Singla, Dr. Sumit Kumar Sonkar, Dr. Vikram Kumar, Dr. Pushpendra Kumar Sharma, Dr. Manish Kumar Chauhan, Mr. Vishnu Agarwal, Dr. Saurabh Sharma, Dr. Jami Paparao, Dr. Digambar Singh, Dr. Krishna Kumar Pandey and Dr. Pragya Berwal Singh.

This book provides an in-depth exploration of e-fuels and their potential to transform the transport sector. The book covers a range of critical topics, including the chemical kinetics of e-fuels, hydrogen production methods and the role of ammonia as a sustainable fuel. It also compares e-fuels with electric vehicles in terms of their effectiveness in reducing emissions. Additionally, the book addresses the technological advancements in hydrogen production, fuel injection techniques and electrofuels, while highlighting the integration of power electronics for efficient fuel production. The chapters include comprehensive analyses, experimental studies and simulation results, offering valuable insights into sustainable fuel solutions for the future of transportation. Through its wide-ranging contributions, this book serves as a vital resource for researchers, engineers and policymakers focused on decarbonizing the transport sector.

Jaipur, India	Nikhil Sharma
Orléans, France	Camille Hespel
New Delhi, India	Snehasish Panigrahy
Agartala, India	Paramvir Singh
Kanpur, India	Avinash Kumar Agarwal

Contents

Editors and Contributors

About the Editors

Dr. Nikhil Sharma is an Assistant Professor at the Department of Mechanical Engineering, Malaviya National Institute of Technology Jaipur, India. He is Lab In-Charge of I.C. Engine Laboratory and is looking after overall management of I.C. Engine Research group. He received his Ph.D. degree from Department of Mechanical Engineering, Indian Institute of Technology Kanpur in 2018. Since 2018, his research has focused on two core areas: emissions from internal combustion engines and optical diagnostics applied to internal combustion engines.

Dr. Camille Hespel is Associate professor at the University of Orleans in PRISME laboratory since September 2009. These main research interests during 15 years are optical diagnostics development, study of spray, droplet vaporization, Diesel and Biofuel combustion, and more particularly the influence of fuel properties on the injection and the impact of air-fuel mixing on the combustion. In this context, she participated in the supervision of five Ph.D. students since 2009 and co-author of 20 publications. She is participating in ADONIS project in EIG CONCERT-Japan (8th Call for Projects in the field of "Sustainable Hydrogen Technology as Affordable and Clean Energy"). She has participated in two national projects: ANR DEVACOL (Diagnostics of vaporization of liquid fuels) and ANR ECN-France (Engine

Combustion Network). She was the initiator and holder of regional project (ECOSIF: 200k€) on the behavior of Diesel injector on cold condition.

Dr. Snehasish Panigrahy is Assistant Professor in the Department of Energy Science and Engineering at the Indian Institute of Technology (IIT) Delhi. Since the beginning of his research career, he has specialized in combustion science, particularly in chemical kinetic modeling for combustion. His expertise encompasses clean fuel technology, ab initio (quantum chemistry) calculations for combustion kinetics, and applications in fluid and thermal engineering. Dr. Panigrahy has established a state-of-the-art rapid compression machine facility at IIT Delhi, the only one of its kind in India to study the auto-ignition properties of various jet and rocket fuels. He has been extensively involved in developing detailed chemical kinetic models to investigate the pyrolysis, ignition, and oxidation of hydrocarbons, ranging from small molecules to heavier and renewable fuels. In addition, Dr. Panigrahy has focused significant efforts on the application of porous medium technology for cost-effective and environmentally sustainable cooking solutions. At IIT Delhi, his Clean Fuel and Energy Laboratory (CFEL) is engaged in various projects funded by organizations such as SERB, ISRO, and Hamilton India, among others. He has also collaborated with global clients, including Siemens Limited, Ford Motor Company, and Milton, contributing to industrial advancements. Dr. Panigrahy has published 28 high-impact papers in renowned journals such as Combustion and Flame, Proceedings of the Combustion Institute, Energy, etc.

Dr. Paramvir Singh currently holds the position of Assistant Professor in the Department of Mechanical Engineering at the National Institute of Technology, Agartala. He completed postdoctoral research in Combustion Chemistry Center at the National University of Ireland, Galway, in 2022, followed by another postdoc focusing on fuels and combustion at the Indian Institute of Technology (IIT), Bombay. In 2019, he was a visiting postdoc in the Mechanical Engineering Department of the University of Sheffield, UK. He earned his Master's and Ph.D. in Mechanical Engineering from the National Institute of Technology Hamirpur, India, in 2013 and 2019, respectively. His research interests span a wide range of topics, including renewable energy, fuel, combustion systems, emission control, gasification, and waste-to-energy technologies. He has a prolific publication record, with more than 50 research articles in international journals and conferences, as well as five book chapters and monographs.

Prof. Avinash Kumar Agarwal is Director at IIT Jodhpur. He obtained his Undergraduate Degree in Mechanical Engineering (1994) from Malaviya Regional Engineering College, Jaipur, and his M.Tech. (Energy, 1996) and Ph.D. (Energy, 1999) from the Indian Institute of Technology (IIT) Delhi. After his Post-Doctoral Fellowship (1999–2001) at the ERC, UW, Madison, USA, he returned to India in 2001 and joined IIT Kanpur. Prof. Agarwal has established a state-of-the-art "Engine Research Laboratory" at IIT Kanpur. Prof. Agarwal is interested in researching IC engines, combustion, conventional and alternative fuels, optical diagnostics, laser ignition, HCCI, particulates and emission control, and large bore engines. Prof. Agarwal has published over 550+ peer-reviewed international journal and conference papers, 65 edited books, and 129 book chapters, attracting 18300+ Scopus and 27400+ Google Scholar citations. For his outstanding contributions, Prof. Agarwal is conferred upon Sir J C Bose National Fellowship (2019) by SERB, Prestigious Shanti Swarup Bhatnagar Prize (2016) in Engineering Sciences, Rajib Goyal Prize in Physical Sciences (2015); NASI-Reliance Industries Platinum Jubilee Award (2012); INAE Silver Jubilee Young Engineer Award (2012); INSA Young Scientist

Award (2007); UICT Young Scientist Award (2007); INAE Young Engineer Award (2005); Devendra Shukla Research Fellowship (2009–12), Poonam and Prabhu Goyal Endowed Chair Professorship (2013–16), SBI Endowed Chair Professorship (2018–21, 2022–25) at IIT Kanpur. He is an elected Fellow of TWAS, SAE, ASME, INAE, ISEES, RSC, NASI, WSSET, AAAS and Combustion Institute USA.

Contributors

Avinash Kumar Agarwal Department of Mechanical Engineering, Indian Institute of Technology Kanpur, Kanpur, Uttar Pradesh, India

Vishnu Agrawal Department of Mechanical Engineering, Malaviya National Institute of Technology Jaipur, Jaipur, Rajasthan, India

Prashant Baredar Energy Centre, Maulana Azad National Institute of Technology, Bhopal, India

Pushpendra Chaturvedi Malaviya National Institute of Technology Jaipur, Jaipur, Rajasthan, India

Ankit Debnath Department of Mechanical Engineering, National Institute of Technology Agartala, Agartala, Tripura, India

Jitendra Dixit Department of Mechanical Engineering, Malaviya National Institute of Technology Jaipur, Jaipur, Rajasthan, India

Gaurav Dwivedi Energy Centre, Maulana Azad National Institute of Technology, Bhopal, India

Pankaj P. Gohil Mechanical Engineering Department, Sarvajanik College of Engineering and Technology, Surat, India

Jai Gopal Gupta Mechanical Engineering Department, National Institute of Technology Sikkim, Ravangla, India

Varun Gupta Electronics and Communication Engineering Department, National Institute of Technology Sikkim, Ravangla, India

Camille Hespel Université d'Orléans, Orléans, France

Geetanjali Kapoor Chandigarh University, Mohali, Punjab, India

Sujeet Kesharvani Energy Centre, Maulana Azad National Institute of Technology, Bhopal, India

Sudarshan Kumar Department of Aerospace Engineering, Indian Institute of Technology Bombay, Mumbai, India

Vivek Kumar Electrical and Electronics Engineering Department, National Institute of Technology Sikkim, Ravangla, India

Krishna Kumar Pandey Department of Mechanical Engineering, National Institute of Technology Agartala, Agartala, Tripura, India

Snehasish Panigrahy Department of Energy Science and Engineering, Indian Institute of Technology Delhi, Delhi, New Delhi, India

Rashmita Patel Chemical Engineering Department, Sarvajanik College of Engineering and Technology, Surat, India

Vaibhav Pathak Department of Energy Science and Engineering, Indian Institute of Technology Delhi, Delhi, New Delhi, India

Nagesh D. Patil Department of Mechanical Engineering, Indian Institute of Technology Bhilai, Durg, India

Sakshi Sarathe Energy Centre, Maulana Azad National Institute of Technology, Bhopal, India

Dilip Sharma Department of Mechanical Engineering, Malaviya National Institute of Technology Jaipur, Jaipur, Rajasthan, India

Nidhi Sharma Amity University Uttar Pradesh, Noida, Uttar Pradesh, India

Nikhil Sharma Department of Mechanical Engineering, Malaviya National Institute of Technology Jaipur, Jaipur, Rajasthan, India

Pravesh Chandra Shukla Department of Mechanical Engineering, Indian Institute of Technology Bhilai, Durg, India

Meenakshi Singhal Guru Nanak Dev University, Amritsar, Punjab, India

Padmini Singh Electrical Engineering Department, Indian Institute of Technology Jammu, Jammu, India

Paramvir Singh Department of Mechanical Engineering, National Institute of Technology Agartala, Agartala, Tripura, India

Sanjay Singh Amity University Uttar Pradesh, Noida, Uttar Pradesh, India

Durgesh Srivastav Discipline of Electric Vehicle Technology, Department of Electrical Engineering, Indian Institute of Technology Bhilai, Durg, India

Sunil Department of Mechanical Engineering, Malaviya National Institute of Technology Jaipur, Jaipur, Rajasthan, India

Suyash Wankhede Department of Mechanical Engineering, Malaviya National Institute of Technology Jaipur, Jaipur, Rajasthan, India

Chapter 1
Introduction to Sustainable, Scalable, and Storable E-Fuels for Decarbonizing Transport Sector

Nikhil Sharma, Camille Hespel, Snehasish Panigrahy, Paramvir Singh, and Avinash Kumar Agarwal

The transport sector is a major contributor to global emissions. This highlights the need for innovative decarbonization solutions to combat climate change and achieve sustainability goals. E-fuels typically derived from renewable energy, water, and captured carbon dioxide, are a promising pathway to sustainability. Such fuels can highly reduce net carbon emissions by reusing CO_2 that has emanated into the atmosphere. Moreover, e-fuels are scalable and storable, making them suitable for the various transport modes, such as aviation, maritime, and heavy-duty road transport, where high energy density and long-range capabilities are vital.

To optimize combustion efficiency, cut emissions, and to support the transition to sustainable sources of energy, understanding chemical kinetic modeling of e-fuels is

N. Sharma (✉)
Department of Mechanical Engineering, Malaviya National Institute of Technology Jaipur, Jaipur, Rajasthan, India
e-mail: nikhil.mech@mnit.ac.in

C. Hespel
Université d'Orléans, Château de la Source, Orléans, France

S. Panigrahy
Department of Energy Science and Engineering, Indian Institute of Technology Delhi, Delhi, New Delhi, India

P. Singh
Department of Mechanical Engineering, National Institute of Technology Agartala, Agartala, Tripura, India

A. K. Agarwal
Department of Mechanical Engineering, Indian Institute of Technology Kanpur, Kanpur, Uttar Pradesh, India

N. Sharma et al. (eds.), *Sustainable, Scalable and Storable E-Fuels for Decarbonising Transport Sector*, Energy, Environment, and Sustainability,
https://doi.org/10.1007/978-981-96-7388-9_1

critical. In this book, Chap. 2 discusses chemical kinetic modeling to study the ignition and oxidation of e-fuels. The global transportation sector relies heavily on traditional fuels, leading to substantial CO_2 emissions. E-fuels, produced using renewable energy, provide a sustainable alternative to mitigate this impact. This chapter explores the pyrolysis and oxidation chemistry of potential e-fuels like methanol, ethanol, and propanol, comparing their ignition characteristics with conventional fuels like methane and ethane. It features reaction path flow diagrams and sensitivity analyses to depict mechanisms of oxidation and identify key reactions that affect performance.

Understanding different hydrogen production methods and their associated emissions is necessary for minimizing the environmental impacts that would make hydrogen no longer a clean energy source. Chapter 3 addresses different hydrogen production methods and associated anthropogenic releases. As hydrogen becomes essential in the clean energy landscape, this chapter reviews the environmental impacts of its production and the emissions involved throughout the hydrogen value chain. It distinguishes between stack emissions from production and fugitive emissions during storage and transportation, discussing technologies like CO_2 capture and storage (CCS) to reduce these emissions and promote sustainable hydrogen production practices.

Understanding how hydrogen is produced and how internal combustion engines are injected is essential to increasing engine performance and lowering greenhouse gas emissions. Methods for producing hydrogen and injecting it into internal combustion engines are covered in Chap. 4. This chapter examines several techniques for producing hydrogen, emphasizing both renewable and fossil fuel-based sources. In order to demonstrate how hydrogen can improve the sustainability of conventional engines, it also examines different hydrogen injection techniques for internal combustion engines, including port fuel injection and direct injection. These techniques are examined to ascertain their effects on emissions, performance, and efficiency.

It is essential to study ammonia as a sustainable fuel in order to analyze its potential to decarbonize energy systems and improve hydrogen storage. Chapter 5 is about the role of ammonia as a sustainable fuel. Ammonia is gaining recognition as a carbon-free fuel with existing infrastructure and high hydrogen density. This chapter discusses various production methods and ammonia's fuel properties, including its potential use in fuel cells and internal combustion engines. It also discusses combustion efficiency and emissions-related issues, which position ammonia in future energy systems.

Hydroxy gas is an alternative to diesel fuel, as it can be used in order to raise the combustion efficiency while minimizing environmental impacts. Chapter 6 discusses about hydroxy gas as diesel fuel substitute. Hydroxy gas also known as HHO is promising substitute for diesel fuel which increase combustion efficiency. This chapter explores the properties and production methods of HHO and its integration into diesel engines, discussing how it could reduce fuel consumption and contribute to sustainable energy development.

Reducing greenhouse gas emissions in the transportation sector and advancing sustainable energy solutions require an understanding of the possibilities of e-fuels and electric vehicles. Electric vehicles and e-fuels for decarbonizing the transportation sector are covered in Chap. 7. The contribution of electric vehicles and e-fuels to reducing greenhouse gas emissions in transportation is contrasted in this chapter. It highlights how both e-fuels and electric vehicles are essential to the transition to a sustainable transportation future while highlighting the advantages and disadvantages of the two options from a technological, financial, and environmental standpoint.

To maximize the efficiency and scalability of renewable energy systems and facilitate the transition to a more sustainable transportation sector, it is important to understand integrated power electronics solutions for e-fuel production. Chapter 8 looks at integrated power electronics solutions for e-fuel production. Innovative approaches are needed to decarbonize transportation, such as the use of sophisticated power electronics in the manufacture of e-fuel. This chapter discusses how these technologies improve the dependability and efficiency of renewable energy conversion processes, which are essential to the production of e-fuels. Successful applications are highlighted in case studies to show how these developments might significantly impact sustainable transportation infrastructure.

To assess their viability as substitutes for traditional fuels and their potential to lower greenhouse gas emissions, electrofuels must be examined as sustainable fuels. Comprehensive analysis of electrofuels as sustainable fuels is the title of Chap. 9. Compared to traditional fuels, electrofuels which are made from renewable energy and CO_2 that has been captured have a significant potential to reduce greenhouse gas emissions. This chapter examines the life cycle effects, production costs, and combustion characteristics of many electrofuel production methods, including power-to-fuel technology. It comes to the conclusion that technological innovation and supportive legislative frameworks are essential for the successful integration of electrofuels.

Maximize gas stoves' energy efficiency and combustion performance to cut down on pollutants and fuel consumption. Thermodynamic properties of an oval shape new flame casing are covered in Chap. 10. Improving home gas stove efficiency is essential for energy conservation. Using ANSYS-CFX simulations, this chapter evaluates the performance of a unique oval flame case and finds that it performs better than conventional burners in terms of maximum combustion temperatures and flame velocities. These results imply that substantial fuel savings and improved stove performance may result from this creative design.

Understanding sustainable smart cities and artificial intelligence is crucial for using technology to improve urban living, optimize resource management, and address environmental challenges. Chapter 11 is about sustainable smart city and artificial intelligence. Smart cities employ technology to tackle urban issues sustainably. This chapter explores how artificial intelligence enhances urban management and facilitates the development of smart cities. Through bibliometric analysis, it identifies research trends and opportunities for integrating AI in sustainable urban planning, emphasizing its importance for future urban development.

In a nutshell, the shift to e-fuels presents a great opportunity for the decarbonization of the transportation sector, which is one of the largest emitters of greenhouse gases globally. E-fuels are a scalable and sustainable alternative to fossil fuels, derived from renewable energy sources and captured CO_2. They have the potential to significantly reduce carbon emissions. However, for them to be widely used, issues including high production costs, energy inefficiency, and the early stage of technological development must be resolved. As the chapters emphasize, maximizing e-fuel efficiency and integration requires a grasp of the chemical processes, developments in hydrogen synthesis, and breakthroughs in engine technology. To achieve a low-carbon future for the transportation industry, e-fuels, electric vehicles, and other sustainable technologies must work together.

Chapter 2
Chemical Kinetic Modeling to Study the Ignition and Oxidation of E-Fuels

Vaibhav Pathak and Snehasish Panigrahy

Abstract The global transportation industry heavily depends on traditional fuel consumption. It accounts for over 50% of total conventional liquid fuels. This reliance has contributed significantly to CO_2 emissions, with transportation alone responsible for nearly a quarter of total CO_2 output. To achieve sustainable transportation, e-fuels present a viable alternative to mitigate the adverse effects of CO_2 emissions from the combustion of fossil fuels, which have been significantly depleted due to excessive usage. E-fuels, known as carbon–neutral synthetic fuels, are produced using renewable electricity for water electrolysis to generate hydrogen and utilize CO_2 from biomass or industrial emissions through the Fischer–Tropsch process, resulting in various e-fuels. Substituting conventional fuels with e-fuels directly reduces emissions from transportation facilities. Hydrogen and alcohol due to their lower carbon content are suitable categories of e-fuels. This chapter focuses on the pyrolysis and oxidation chemistry of potential e-fuels, including methanol, ethanol, and propanol. It examines their ignition delay time characteristics. These characteristics are compared to methane, ethane, propane, ethene, and propene. The goal is to assess the suitability of these e-fuels for use in combustion chambers. Additionally, it includes reaction path flow diagrams and sensitivity analysis to understand the oxidation chemistry and identify chain-promoting and inhibiting reactions in the mechanism. It provides an introduction and overview of alcohol and alkane combustion by modeling the results with robust chemical kinetic mechanisms.

Keywords Conventional fuels · CO_2 emissions · Fischer–Tropsch · Hydrogen · e-fuels · Alcohol

V. Pathak · S. Panigrahy (✉)
Department of Energy Science and Engineering, Indian Institute of Technology Delhi, Delhi, New Delhi, India
e-mail: Snehasish.Panigrahy@dese.iitd.ac.in

N. Sharma et al. (eds.), *Sustainable, Scalable and Storable E-Fuels for Decarbonising Transport Sector*, Energy, Environment, and Sustainability,
https://doi.org/10.1007/978-981-96-7388-9_2

2.1 Introduction to E-Fuels

Global warming represents a critical threat to our planet, primarily driven by the continuous rise in greenhouse gas emissions. This escalating release of gases, notably CO_2, elevates global temperatures by trapping heat in the Earth's atmosphere. The resultant warming disrupts weather patterns, placing ecosystems, agriculture, and human health under unprecedented threat from more frequent and intense heatwaves, storms, and droughts. The transportation industry plays a pivotal role in this scenario, being a major contributor to both CO_2 emissions and global energy consumption, responsible for approximately 25% of CO_2 emissions and consuming over half of all produced liquid hydrocarbons (Ramirez et al. 2020). Given that these hydrocarbons comprise over 95% of the sector's energy use, the urgency for transitioning to low-carbon or carbon–neutral alternatives is evident (Chehrmonavari et al. 2023). Internal combustion engine vehicles, which represent about 99.8% of the transportation sector, consume 25% of the oil-based energy globally. The International Energy Agency reported in 2023 that global CO_2 emissions surged by 410 million tons (nearly 1.1%) from the previous year, reaching an all-time high of 37.4 billion tons (IEA 2024). In addition to CO_2, vehicle exhausts emit harmful nitrogen oxides (NOx) and particulate matter, which further harm the environment (Frost et al. 2023). Therefore, the transportation sector is a critical focus for reducing energy consumption and carbon emissions, accounting for more than 22% of all greenhouse gas (GHG) emissions and nearly 26% of the world's total energy demand. According to the Paris Agreement, a significant global reduction in carbon emissions is necessary to achieve zero emissions between 2050 and 2060 and limit global temperature rises to 2 °C (Huey et al. 2024). To align with the Net Zero Emissions Scenario, transport emissions must decrease by about 25% to 6 Gt by 2030, as illustrated in Fig. 2.1, the data for which has been taken from the International Energy Agency (IEA 2021).

Fuel transformation is crucial as internal combustion engines (ICEs) remain necessary despite advances in alternative power sources. E-fuels, or electro-fuels, are synthetic fuels which are produced by reacting green hydrogen (H_2) generated from electrolyzing water using renewable energy with absorbed and captured CO_2. This process provides a practical replacement for conventional fossil fuels, utilizing existing infrastructure and significantly reducing greenhouse gas emissions (Ram and Salkuti 2023) (Fig. 2.2). E-fuels like e-methanol offer numerous environmental benefits, including availability, minimal energy storage requirements, large-scale production potential, and ease of transport. Due to their low environmental impact and potential to reduce reliance on traditional fossil fuels, e-fuels are becoming the preferred fuel source, ensuring a sustainable future for internal combustion engines. Notably, the use of renewable energy in hydrogen production is a pivotal stage in e-fuel production. Water electrolysis, powered by renewable sources like solar or wind energy, is a method of splitting water into hydrogen and oxygen with minimal greenhouse gas emissions. The resulting hydrogen, known as e-hydrogen, is integral to the synthesis of e-fuels such as e-methanol and e-ammonia. Carbon capture, utilization, and storage also play a vital role. Direct air capture technologies are employed to

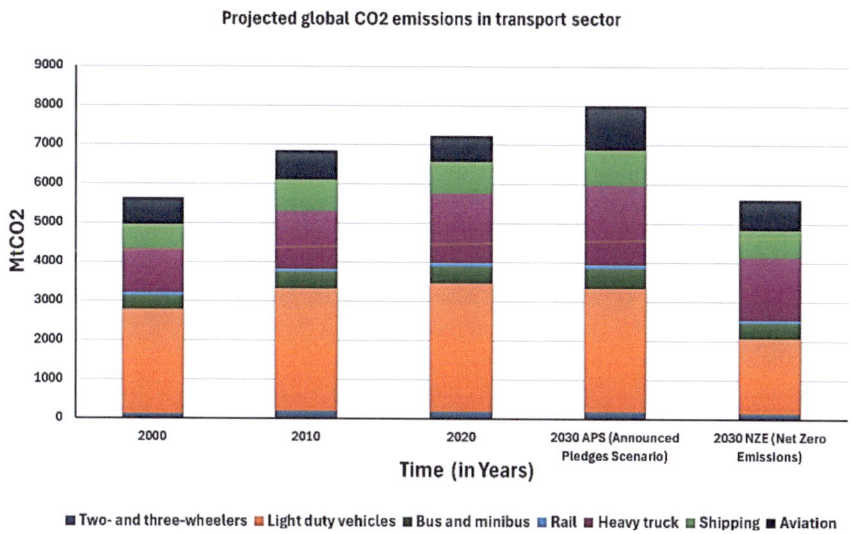

Fig. 2.1 Projected global CO_2 emissions from different vehicles in the transport sector (IEA 2021)

capture CO_2 directly from the air, while pre-combustion, post-combustion, and oxy-combustion methods are used to capture CO_2 from industrial emissions (Shi et al. 2024).

E-fuels, or synthetic fuels, are produced following a number of stages that involve capturing carbon dioxide (CO_2) either from power plants or directly from the air, generating hydrogen from renewable electricity using electrolysis, and synthesizing the final fuel product through the Fischer–Tropsch process. Each stage presents unique challenges and efficiencies that are critical to understanding the overall viability of e-fuel production.

Fig. 2.2 Schematic representing input requirements for e-fuel production

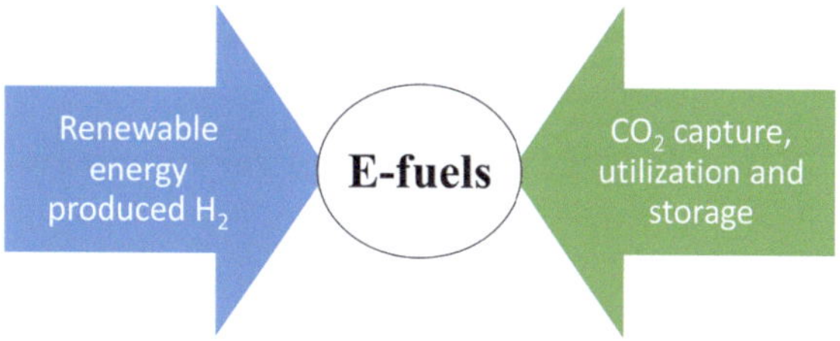

2.1.1 Carbon Capture in E-Fuel Production

The first step in synthesizing e-fuels is capturing carbon dioxide (CO_2) either from industrial emissions summarized as point source capture or from the atmosphere using direct air capture (DAC). There are two primary DAC technologies:

Solvent-based DAC: Air is passed through a chemical solvent that binds to CO_2. The CO_2 is then separated by applying heat or a vacuum to release it from the solvent, allowing cleaned air to be released back into the atmosphere.

Sorbent-based DAC: Solid materials, or sorbents, bind with CO_2 from the air. When heated or placed under a vacuum, the CO_2 is released in concentrated form for capture or storage.

The low atmospheric CO_2 content (around 0.04%) presents serious difficulties for both DAC techniques. Higher energy needs and operating expenses result from the difficulty of extracting CO_2 from the air as opposed to capturing it from high-concentration point sources like power plants. The expense of CO_2 capture is still a significant problem; although industrial emissions are less expensive to capture because of their higher concentrations, DAC frequently costs more than $100 per ton of CO_2. Lowering DAC costs and increasing scalability are the goals of programs like the Department of Energy's (DOE) Carbon Negative Shot.

Although there are still obstacles to overcome, DAC technologies are developing quickly. In both solvent-based and sorbent-based systems, researchers are working to increase efficiency while lowering costs. Better CO_2 conversion or storage techniques, integration with renewable energy sources, and improvements in chemical processes are important areas of advancement. The scalability of DAC is essential; these technologies have the potential to greatly aid in the worldwide effort to reduce carbon emissions if they can be implemented on a bigger scale and incorporated into the manufacture of e-fuel. In conclusion, although if DAC technologies have financial and energy constraints, they are essential to the generation of e-fuel. Making e-fuels a practical choice for cutting emissions in sectors that are challenging to electrify will require continued advancements in carbon capture technology.

2.1.2 Hydrogen Production via Electrolysis

One of the main methods for producing hydrogen is electrolysis, which will help the Hydrogen Energy Earthshot target of 80% clean hydrogen cost reduction to $1 per kilogram in ten years ("1 1 1"). Electrolysis can produce hydrogen with minimal greenhouse gas emissions, depending on where the electricity comes from. The cost and efficiency of the necessary electricity, as well as emissions from electricity generation, must be taken into account when assessing the advantages and economic feasibility of electrolysis-based hydrogen production.

Due to the greenhouse gases emitted and the quantity of fuel needed due to the low efficiency of the electricity generating process, the current power grid is not suitable for supplying the electricity needed for electrolysis in many parts of the nation. For nuclear and renewable energy sources (wind, solar, hydro, and geothermal), electrolysis is being used to produce hydrogen. The production cost must be drastically reduced to be competitive with more established carbon-based pathways like natural gas reforming, even if these hydrogen production pathways produce almost zero emissions of greenhouse gases and other environmental pollutants.

2.1.3 Synthesis Process

Using techniques like Fischer–Tropsch synthesis or methanol synthesis, collected CO_2 is combined with hydrogen to create e-fuels. A variety of hydrocarbons appropriate for a range of uses can be produced at this stage. To optimize yield and reduce byproducts, however, careful control of reaction parameters (temperature, pressure) and catalyst selection are needed. The quality of the hydrogen generated and the efficacy of CO_2 collecting techniques both affect how efficient this process is going to be.

E-fuels have widespread applications across various sectors, offering a sustainable alternative to traditional fuels. When utilized as Sustainable Aviation Fuel (SAF), they help reduce the carbon footprint of air travel. E-fuels such as ammonia and methanol are environmentally friendly substitutes for heavy fuel oil in marine transportation, minimizing the negative impact on the environment. In heavy-duty transportation, trucks and buses can effectively run on e-fuels, particularly in situations where electrification is not feasible due to weight and range constraints. As electric vehicles (EVs) gain popularity, e-fuels serve as a transitional option for passenger cars, mitigating emissions without requiring immediate changes to the infrastructure. E-fuels possess several advantages over other alternative fuels. They require minimal modifications as they can be easily integrated into the existing fuel infrastructure. With their high energy density, they are suitable for applications requiring high power and long range. Additionally, when produced using captured CO_2 and renewable energy, e-fuels can achieve a net-zero carbon impact by offsetting the CO_2 released during combustion with the CO_2 captured during production. Promoting their use and ensuring their economic competitiveness necessitates substantial government support, including carbon pricing and subsidy programs.

2.2 Fundamentals of Chemical Kinetic Modeling

Combustion is widely regarded as one of the most complex and interdisciplinary fields of study, demanding a deep understanding across a broad spectrum of scientific disciplines. It requires a thorough grasp of chemistry to understand the intricate molecular reactions that drive combustion; physics to comprehend the physical principles that govern the behavior of gases and flames; fluid dynamics to analyze the flow of gases and liquids within combustion systems; thermodynamics to study the energy transformations that occur during combustion; mathematics, to model and predict combustion processes; and computer science, to simulate and analyze these processes using advanced computational tools (Ajulu 2020). This integration of diverse fields is not only essential for advancing the scientific understanding of combustion but also for addressing some of the most pressing environmental issues facing society today, particularly those related to air pollution and the efficient use of energy resources.

The importance of combustion research in society cannot be overstated. The generation of air pollutants is a significant concern, and the ability to study and analyze combustion processes is critical to develop solutions that can reduce harmful emissions. To properly understand the combustion of a fuel, one must comprehend multiple layers of information, starting at the molecular level as the base and extending all the way to the practical application of this knowledge to design the modern combustors. These combustors are engineered to operate with increased efficiency and reduced emissions, aligning with the global imperative to create more sustainable and environmentally friendly energy solutions (Curran 2019). Figure 2.3 presents a diagram that illustrates the steps involved in tackling a combustion problem and shows how a deep understanding of molecular interactions can lead to the development of advanced combustion technologies.

Combustion research is generally divided into four broad levels of development; these levels follow a hierarchy where each level builds upon the knowledge and discoveries of the previous one. The first level involves quantum mechanics and

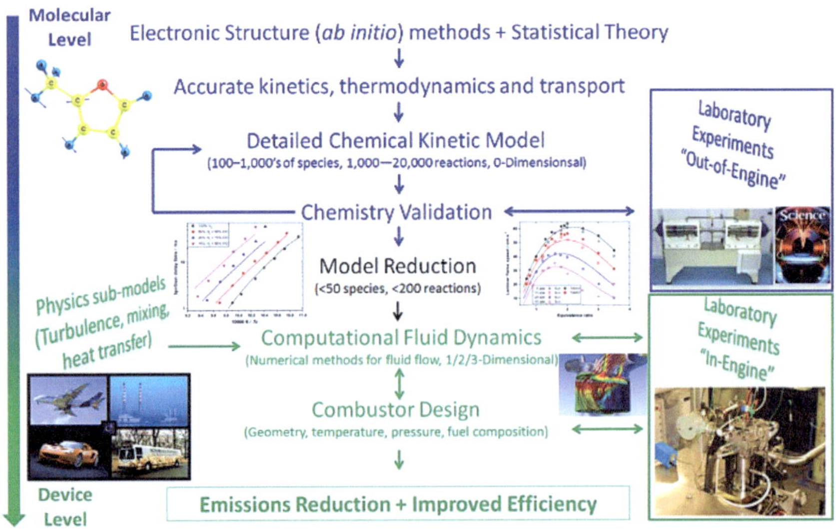

Fig. 2.3 Schematic diagram illustrating the various levels of understanding combustion, from theoretical concepts to practical applications (reprinted from Curran (2019) with permission of Elsevier)

the direct kinetic measurement of rate constants, as well as the identification of reaction intermediates and products. This level is highly theoretical and focuses on the fundamental interactions between molecules at the atomic level. Researchers at this stage use advanced techniques such as ab initio methods and statistical theory in order to determine thermochemical parameters and rate constants with higher level of accuracy. These calculations are essential for understanding the basic building blocks of combustion processes, providing the foundational data needed to model more complex systems (Curran 2019).

Moving to the second level, the focus shifts to fuel structure and fundamental chemistry. Here, the species and reactions identified in the first level are integrated into detailed chemical kinetic mechanisms. These mechanisms are designed to describe the complete reaction pathways that occur during combustion, from the initial breakdown of fuel molecules to the formation of final products. The development of these mechanisms is an iterative process that involves extensive validation against experimental data. Researchers begin with simple systems, such as homogeneous reactors and laminar flames, where conditions are easier to control and measure. As the mechanisms are refined, they are used to predict the experimental data which have been taken in the fundamental experimental setups like rapid compression machines (RCM), high-pressure shock tube (HPST) which more closely resemble real-world conditions without considering the effects of the fluid dynamics over the treatment of the mixture of fuel and oxidiser in the combustion chamber. This level of research is crucial because it bridges the gap between fundamental chemistry and practical applications, ensuring that the theoretical models developed in the first level are accurate and applicable to real-world combustion systems (Curran 2019).

The third level of development is where computational fluid dynamics (CFD) studies come into play. At this stage, the detailed chemical kinetic mechanisms developed in the second level are simplified to create reduced mechanisms that retain the essential features needed for practical applications. These features might include key performance metrics which are determining ignition delay time characteristics, flame speed, and emissions predictions. The reduction process involves carefully selecting the most critical reactions and species, simplifying the mechanisms without sacrificing accuracy. These include sensitivity analysis following the different methods named as Rabitz et al. (1983), Turányi (1990), Directed Related Graph Method (DRG) (Lu and Law 2005), Directed Related Graph with Error Propagation (DRGEP) (Pepiot-Desjardins and Pitsch 2008), Directed Related Graph with Sensitivity Analysis (DRGASA) (Sankaran et al. 2007; Zheng et al. 2007), Directed Related Graph with Error Propagation and Sensitivity Analysis (DRGEPSA) (Niemeyer et al. 2010), Computational Singular Perturbation (Massias et al. 1999), and Rate-Controlled Reduced Equilibrium methods (RCCE) (Keck and Gillespie 1971). These reduced mechanisms are then used in CFD simulations to model the behavior of combustors under various operating conditions. For instance, the NUIGMech1.1 mechanism is intended to simulate the pyrolysis and oxidation of different C0-C4 fuels. It has 771 species and 3,783 reactions. In the work presented by Wang et al. (2021), NUIGMech1.1 mechanism is made simpler by creating basic skeletal mechanisms for the principal fuel species like CH_4, CH_3OH, C_2H_6, C_2H_4, C_2H_5OH, C_3H_8, and many others using a mixture of skeletal reduction approaches. Sampling reaction state points from autoignition simulations at various temperature, pressure, and equivalency ratio conditions are part of the reduction process. Following a brute-force sensitivity analysis to make sure the resulting basic skeletal mechanisms maintain an ignition delay time error of less than 10%, the skeletal mechanisms are further modified using the directed relation graph (DRG) and DRG with error propagation (DRGEP) approaches. The reduced mechanisms are then applied to predict combustion characteristics with high accuracy as mentioned in that paper.

The reduced or skeletal mechanism produced from the detailed mechanism using appropriate reduction techniques mentioned above will reduce the computational cost to run the mechanism and test the system for fuel of interest. The work published by Strelkova et al. (2008) proposed a skeletal mechanism for Jet A fuel so that the combustion characteristics for that fuel can be studied in a gas turbine combustor. They proposed a three-species surrogate mixture, composed of 72.7 wt% decane, 9.1 wt% hexane, and 18.2 wt% benzene, to represent Jet A fuel and made a detailed mechanism of 417 reversible reactions and 71 components. The performance of the detailed mechanism has been validated against the experimental data for Jet A fuel at pressures of 1 and 10 atm, covering lean, stoichiometric, and rich Jet A/air mixtures (equivalence ratios of 0.5, 1, and 2) within the temperature range of 1000–1700 K. A reduced mechanism for Jet A surrogate combustion was derived from the detailed mechanism and implemented in CFD simulations of gas turbine combustors to reduce the computational cost. This reduced mechanism consists of 38 reactions and 24 species and applies over various temperatures, pressures, and equivalence ratios. The results from the reduced mechanism were found to be in reasonable agreement with the detailed mechanism, which in turn showed good correlation with experimental measurements.

In another study by Marques et al. (2020), a reduced mechanism of ethanol was developed so that the combustion characteristics for ethanol could be evaluated for combustion under ultra-lean conditions in spark-assisted homogeneous charge compression ignition (HCCI) rotary engine via CFD simulations. The study concluded that the skeletal mechanism with 188 reactions predicts ignition delay times and laminar flame speeds accurately under ultra-lean ethanol/air conditions at high pressure. The authors have suggested that the engine tests should be executed at an rpm of 2500 with a boost pressure of 2 bar for the CFD model calibration using the optimized reaction mechanism.

The insights gained from these simulations are invaluable for designing new, cleaner, and more efficient combustion systems. This level of research is where the detailed chemical knowledge from the earlier stages is translated into practical engineering tools that can be used to optimize the design and operation of combustion devices (Curran 2019).

Finally, the fourth level of combustion research focuses on practical applications. Researchers working at this level study real-world systems such as jet engines, diesel engines, natural gas safety systems, and fuel inhibition techniques. This applied research is crucial for translating the scientific advancements made at the earlier levels into tangible improvements in combustion technology. It involves close collaboration with industry and regulatory bodies to ensure that new technologies meet the necessary safety, efficiency, and emissions standards. The goal at this stage is to take a detailed understanding of combustion processes developed in the lab and apply it to the design and optimization of systems that are used in everyday life. This might involve developing new fuel combinations, improving the performance of existing engines, or designing entirely new types of combustors that operate with lower emissions and higher efficiency.

The field of chemical kinetic modeling, crucial to combustion research, has advanced significantly with the development of critical mechanisms like GRI-Mech and the primary reference fuel (PRF) mechanisms from Lawrence Livermore National Laboratory. GRI-Mech, a detailed model for natural gas combustion, became one of the first widely accessible online tools, offering a consistent framework for researchers globally and aiding in the accurate prediction of NOx emissions. Similarly, the PRF mechanisms, which model the combustion of n-C_7H_{16} and i-C_8H_{18}—representative of gasoline and diesel—provided essential insights into the combustion processes of these common fuels. The availability of these models marked significant progress in combustion research, facilitating consistent comparison and validation across various studies (Curran 2019).

In recent years, researchers have continued to build on these foundational mechanisms, developing new models that describe the combustion of larger hydrocarbon molecules and oxygenated hydrocarbons. These advancements have been significant for studying the combustion of alternative fuels, which are increasingly being used in an effort to mitigate harmful emissions responsible for global warming and our dependence over conventional fuels. The step-by-step development of surrogate mechanisms that accurately represent the combustion of complex fuels has been a major focus of recent research, and significant progress has been made in this area.

By understanding and improving these fundamental processes, combustion researchers can continue to push the boundaries of what is possible in the quest for cleaner, more efficient energy solutions. The advancements made in chemical kinetic modeling have the potential to revolutionize the way we think about and use energy, providing new pathways to sustainability and helping to cope with some of the critical environmental-prone problems of our time. As the field continues to evolve, the integration of interdisciplinary knowledge will remain crucial, ensuring that combustion science stays at the forefront of technological innovation and environmental stewardship.

A chemical kinetic mechanism is a numerical model that comprehensively includes the detailed elementary reactions of the reacting species, thermodynamic estimates, corresponding rate constants involving collider species, and the transport properties of species under specific reactor conditions of temperature, pressure, and mole fractions. To understand this better, we can take up an example for modeling the H_2–O_2 system; up to 40 reactions may be considered, which involve the following eight species: $H_2/O_2/O/H/OH/HO_2/H_2O$ and H_2O_2 (Stephen 2000). Each reaction at an elementary level emphasizes the reactant and product species, along with the rate constant for both directions, whether forward or reverse, in case of a reversible reaction.

$$\mathrm{R} \underset{k_r}{\overset{k_f}{\rightleftharpoons}} \mathrm{P}. \tag{2.1}$$

The forward rate constant k_f is commonly denoted as k_f [R], while the reverse rate constant k_r is typically expressed as k_r[P]. This reaction rate constant is determined by the Arrhenius equation, which is defined as:

$$k_f = A \times e^{-\dfrac{E_a}{RT}} \tag{2.2}$$

Or that of the modified Arrhenius

$$k_f = A \times T^n \times e^{-\dfrac{E_a}{RT}}. \tag{2.3}$$

At equilibrium, the rate constants for forward and reverse reactions are equal, thus

$$K_{\text{eq}} = K_c \tag{2.4}$$

Also, that

$$K_{\text{eq}} = -\frac{\Delta G_r}{RT} \tag{2.5}$$

And that

$$K_c = \frac{k_f}{k_{\mathrm{r}}}. \tag{2.6}$$

Then, at chemical equilibrium,

$$K_{\text{eq}} = K_c = \frac{k_f}{k_r}, K_{\text{eq}} = \frac{k_f}{k_r} \tag{2.7}$$

And knowing that

$$\Delta G_r = \Delta H_R - T \Delta S_R, \tag{2.8}$$

where ΔG_r represents the Gibbs free energy of the reaction; it signifies the energy available for performing non-boundary (PV) work in a thermodynamically closed system at constant temperature and pressure. A negative ΔG_r corresponds to a chemical reaction in which the formation of products is favored, while a positive value suggests reactants are favored. Thus, it determines the direction of the reaction at particular temperature and pressure conditions. ΔH_R is the enthalpy of the reaction, signifying the nature of the reaction to be either exothermic or endothermic, and ΔS_R is the entropy of the reaction. Therefore, once the thermo-kinetic parameters of the forward rate constant are known for reacting species, the corresponding reverse rate constant can be determined using the equilibrium constant, as thermodynamic equilibrium is achieved. A chemical kinetic mechanism includes the thermo-kinetic parameters for the forward rate constants of elementary reactions at specified temperatures and pressures, along with thermodynamic estimates and third-body colliders. Additionally, the mechanism incorporates transport properties to determine flammability characteristics.

Reaction rates and equilibrium constants mentioned above corresponding to the elementary reactions participating in the mechanism to predict combustion characteristics are analyzed to optimize the performance of combustion systems, eliminating safety concerns while designing the combustion chamber, as the pressure and temperature conditions for the particular mixtures are already predicted in CFD. The hierarchy usually followed to merge the benefits of using chemical kinetics and computational fluid dynamics applications has been illustrated schematically in Fig. 2.3

2.3 Simulation Techniques and Tools

The connection between chemical kinetic mechanisms and computing power has grown substantially over the years, making it nearly impossible to develop models without the support of these numerical networks. With advancements in computing technology and the refinement of numerical methods, more models are now generated using automated software and interactive algorithms for simulating chemical mechanisms. This is evidenced by the rise of tools such as reaction design's CHEMKIN suite—initially developed by Sandia National Laboratories in the 1980s and now owned by ANSYS (https://www.ansys.com/products/fluids/ansys-chemkin-pro)—along with other software like EXGAS (Warth et al. 2000), OPENSMOKE++ (Cuoci et al. 2013, 2015; Stagni et al. 2014), LOGE-soft (https://logesoft.com), Cantera (Goodwin et al. 2018), FlameMaster (Pitsch 1990), CosiLab (Razus et al. 2022), CMCL Innovations kinetics (http://cmclinnovations.com/solutions/products/kinetics/), DETCHEM (2004), Chemical WorkBench (http://www.kintechlab.com/products/chemical-workbench/), and RMG (Gao et al. 2016; Allen et al. 2012; Magoon and Green 2013). These software tools are proficient in generating chemical kinetic mechanisms and validating them against experimental data.

For effectively simulating chemical species, automatic chemical kinetic mechanism generators must possess four key capabilities. Firstly, they need a reliable method to accurately identify and represent chemical species. Secondly, they require a systematic approach to determine possible reactions and how they occur between reactive species. Thirdly, they must have a way to obtain accurate thermodynamic and kinetic parameter estimates. Lastly, there must be criteria for deciding which species to include or exclude in the reaction model.

These numerical models continuously solve differential equations for chemically reactive species based on initial conditions, often derived from experiments or boundary conditions. The integration of these species is conducted by applying the conservation laws of energy, mass, species, and linear momentum during flame simulations within a defined step size. As the integration progresses, the time steps are adjusted to estimate quantities like temperature, pressure, velocity, and species concentrations or mole fractions. The methods employed by most numerical solvers depend on the physical system being modeled, the geometry, the step size, and the imposed convergence criteria. If the time step changes too rapidly for the solver scheme, it can lead to failure or inaccurate approximations. Therefore, most solvers are typically restricted to using uniform step sizes and maintaining constant parameters to achieve more efficient computation.

After learning everything about the different tools and techniques involved in studying and analyzing the combustion characteristics of the fuel of interest, let us examine its application in the next section.

2.4 A Comparative Study of Autoignition Properties of Saturated and Unsaturated Hydrocarbon and Alcohols Ranging from C_1–C_3

C3MechV3.3 is employed in this study to simulate CH_4, C_2H_6, C_3H_8, C_2H_4, C_3H_6, CH_3OH, C_2H_5OH, nC_3H_7OH, and iC_3H_7OH, developed using current kinetic models alongside the latest advancements and expertise from the C3 research groups. The Computational Chemistry Consortium (C3) has introduced C3MechV3.3. This mechanism, based on a C0–C4 core, focuses on complicated surrogates of long chain hydrocarbon fuels like *n*-hexane and its corresponding isomers, *n*-heptane, iso-octane, *n*-C_8–*n*-C_{12} saturated hydrocarbons, polycyclic aromatic hydrocarbons (PAHs), and NOx as key species of interest. The kinetic model is composed of the most recent chemical subsets developed by various C3 partners.

Ignition delay time characteristics data is measured from high-pressure shock tube (HPST). It is defined as the time interval between the pressure rise due to the shock arrival at the end wall and the sharp pressure rise due to ignition as shown in, Fig. 2.4 (Mohamed et al. 2022). Along with IDT, decomposition and production of species mole fraction, sensitivity, and production rate analyses, the Homogeneous Closed Volume Reactor module in CHEMKIN-Pro is used. The chemical kinetic models were then validated through zero-dimensional modeling, incorporating the mixture composition, initial compression pressure, and temperature as inputs. In zero-dimensional modeling, the system is represented as a single control volume or reactor without accounting for spatial variations, hence referred to as a lumped volume. The model assumes the system is uniformly mixed, maintaining constant internal temperature, pressure, and composition. This simplification allows for a more straightforward mathematical representation of the system's behavior. The

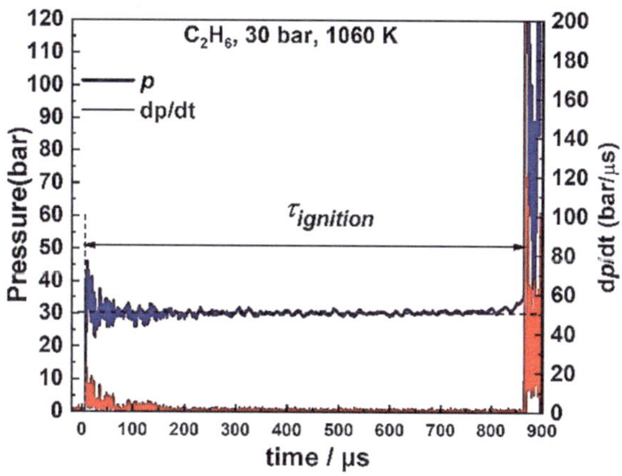

Fig. 2.4 Ignition delay time (IDT) data determined from pressure–time histories of high-pressure shock tube (HPST) data are illustrated for the C_2H_6 at 30 bar and 1060 K (reprinted from Mohamed et al. (2022) with permission of Elsevier)

governing equations include the conservation of mass, energy, and species, describing the energy balance within the reactor and the rate at which species concentrations change. Reaction rates are determined using rate expressions and constants derived from experimental data or kinetic models.

Figure 2.5 illustrates newly simulated ignition delay times (IDT) characteristics using the CHEMKIN-Pro software with the C3MechV3.3 mechanism file and the IDT data reported by Martinez et al. (2021). This comparison involves CH_4, C_2H_4, C_2H_6, and C_3H_8 mixtures across various concentrations, pressures, temperature ranges, and stoichiometric conditions.

Figure 2.6 displays the ignition delay times for various alcohols and hydrocarbon fuels under a specific temperature and pressure. The graph highlights several important trends, particularly that methane exhibits the slowest ignition among the alkane fuels for compression temperature range of 625–1700 K. In the lower temperature range, methane's reactivity is reduced due to the competition for the CH_3O_2 radical to form methoxy radicals (CH_3O) by $CH_3\dot{O}_2 + O_2 \rightleftharpoons CH_3\dot{O} + CH_3\dot{O}$ and $CH_3\dot{O}_2 + CH_3\dot{O}_2 \rightleftharpoons 2CH_3\dot{O} + O_2$ reactions, and the other is by chain-propagating reactions $CH_2O + CH_3\dot{O}_2 \rightleftharpoons H\dot{C}O + CH_3OOH$ and $CH_4 + CH_3\dot{O}_2 \rightleftharpoons CH_3OOH + \dot{C}H_3$.

$H + O_2 \rightleftharpoons \ddot{O} + \dot{O}H$, a chain-propagating reaction usually drives and improves the fuel/Air (F/A) reactivity combinations in the high-temperature range. Hydrogen radicals ($\dot{H}$) are not as important in the ignition process of a CH_4/air mixture as methyl radicals ($\dot{C}H_3$) are in ignition. Methane has the slowest rate of ignition among the alkanes due to a significant amount of these $\dot{C}H_3$ radicals undergoing self-recombination to produce C_2H_6.

The main process driving the consumption of ethane is H-abstraction, which produces an ethyl radical ($\dot{C}_2H_5$), which then decomposes into a $\dot{H}$ radical and ethene (C_2H_4). The resultant $\dot{H}$ radical increases reactivity of the mechanism via a chain-branching reaction in which it reacts with molecular oxygen and forming oxygen

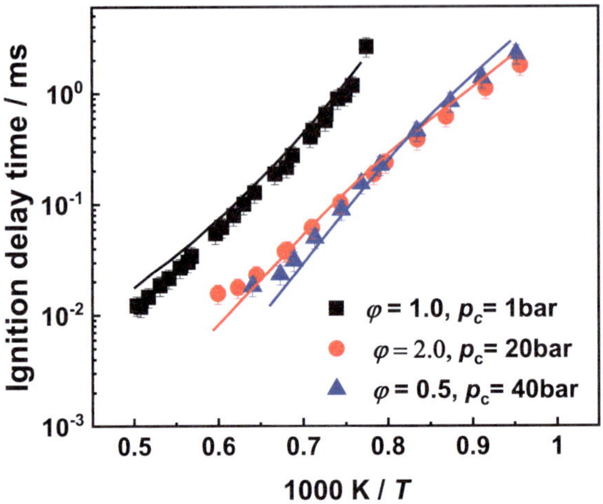

Fig. 2.5 Ignition delay time (IDT) data from simulations and experiments are illustrated for the mixtures of $CH_4/C_2H_4/C_2H_6/C_3H_8$. A 60% CH_4/15% C_2H_4/15% C_2H_6/10% C_3H_8 blend is included in the data, along with three distinct dilutions: 75% N_2 (shown by red circles/lines), 75% N_2 + 15% Ar (represented by black squares/lines), and 85% N_2 (represented by blue triangles/lines). Experimental IDT data is reported by Martinez et al. (2021)

Fig. 2.6 Ignition delay time (IDT) measurements based on the C3MechV3.3 mechanism predictions were conducted under stoichiometric conditions with primary *n*-alkanes and alkenes from C_1–C_3 $CH_4/C_2H_4/C_2H_6$ and methanol (CH_3OH) and ethanol (C_2H_5OH) combined with air at a pressure of 50 bar (reprinted from Pathak and Panigrahy (2024) with permission of Begell House, Inc.)

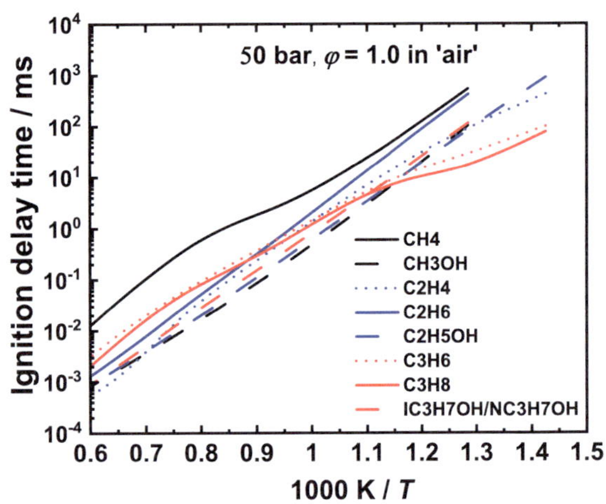

($\ddot{O}$) and hydroxyl radicals ($\dot{O}H$), $\dot{H} + O_2 \rightleftharpoons \ddot{O} + \dot{O}H$. The analysis of reaction flow pathway of C_2H_6 at a high temperature of 1400 K, as seen in Fig. 2.7b, reveals that the ethyl radical combines with O_2 and forms ethene (C_2H_4) by undergoing H-abstraction process. But this reaction hinders overall reactivity since it competes with the ethyl radical's breakdown.

Propane has higher reactivity than ethane and methane at low temperatures due to a combination of variables. Propane has more carbon atoms and a larger molecular

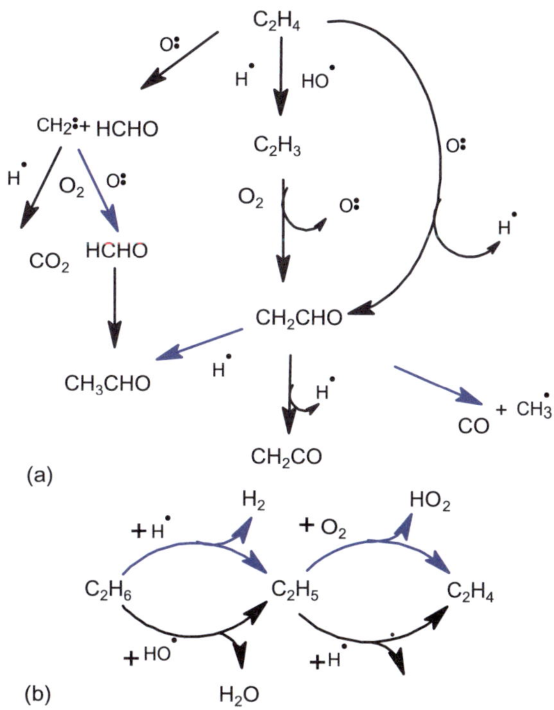

Fig. 2.7 Flux analyses of pure **a** C_2H_4, **b** C_2H_6 fuel ignition for compression temperature and pressure of 1350 K and 50 bar respectively and 18% fuel consumption (blue arrow representing the chain inhibiting reactions while black arrows indicate chain-propagating reactions) (reprinted from Pathak and Panigrahy (2024) with permission of Begell House, Inc.)

structure, which is more reactive overall because reactive propyl radicals (C_3H_7) are formed more quickly and engage in additional processes. The stability of the alkyl depends upon the stability of the alkane from which it is formed, either by H-abstraction, pyrolysis or by undergoing another method. The stability of alkane decreases as the number of carbons in the chain increases on account of the potential energy surface, indicating that propane is less stable than methane and ethane and hence propyl radicals will be formed more quickly. Propane is also more prone to undergo hydrogen abstraction by reactive species due to its higher hydrogen atom count compared to ethane and methane. Reactivity is increased much more by this start of radical chain reactions. Propane's molecular structure and extra carbon atoms also cause the carbon–carbon bonds to be slightly lessen in strength than those in ethane and methane. As a result of the reduced binding strength, subsequent reactions are made easier to occur (Pathak and Panigrahy 2024).

Furthermore, propane oxidation is predominantly fueled by H-atom abstraction by $OH^{\cdot}$ radicals and $H^{\cdot}$ atoms at high temperatures, resulting in the generation of primary ($n\dot{C}_3H_7$) and secondary propyl ($i\dot{C}_3H_7$) radicals. Ethyl ($\dot{C}_2H_5$) and methyl ($\dot{C}H_3$ radicals are created when propane undergoes unimolecular breakdown at these high temperatures. While methyl radicals are involved in a reaction that reduces reactivity ($\dot{C}H_3 + H\dot{O}_2 \rightleftarrows CH_4 + O_2$), they also participate in a chain-branching reaction ($\dot{C}H_3 + H\dot{O}_2 \rightleftarrows \dot{C}H_3O + \dot{O}H$), which increases reactivity. Propane's ignition delay time characteristics are influenced by the competition between these two channels. Propane's reactivity is further decreased at high temperatures by the substantial recombination of CH_3 radicals to create C_2H_6.

Furthermore, due to the elevated H_2O_2 generation, propane displays a negative temperature coefficient (NTC) region. Since H_2O_2 is thermally stable at low and intermediate temperatures, its reactivity decreases. Nevertheless, H_2O_2 quickly breaks down into two reactive $\dot{H}$ radicals when the temperature climbs high enough, greatly increasing reactivity.

However, more reactivity indicated by the shorter ignition delays of the alkenes propene and ethene is exhibited as compared to their respective alkanes. At 1430 K, hydroxyl radicals ($\dot{O}H$) and hydrogen radicals ($\dot{H}$) abstract hydrogen atoms from ethene, resulting in the production of vinyl radicals (C_2H_3). Following a chain-branching reaction, these vinyl radicals combine with O_2 and generate vinoxy radicals ($\dot{C}H_2CHO$). By generating hydrogen radicals through reactions with ethene, oxygen atoms further increase reactivity. They also produce $\dot{C}H_2CHO$ and $\dot{H}$, as well as $\dot{C}H_2$ and formaldehyde (CH_2O) through other pathways.

The hydrogen radicals ($\dot{H}$) produced participates aggressively in chain-branching reactions, where they interact with O_2 to form hydroxyl radicals ($\dot{O}H$) and $\ddot{O}$ radicals. At significantly higher temperatures, however, ethane is less reactive than ethene because of competing reactions involving ethyl radicals ($\dot{C}_2H_5$) and O_2, which result in ethene (C_2H_4) and hydroperoxyl radicals ($H\dot{O}_2$), as well as hydrogen atom abstraction by hydrogen radicals ($\dot{H}$) from the ethane.

The chemical structure of propene has an allylic site, which makes it less reactive at high temperatures. About 12% of propene is consumed as methyl-vinyl radicals are created by allylic reactions, which lowers the flux toward chain branching. In addition, H_2 at the allylic site consume about 7% of propene by H-atom abstraction. The resultant H_2O_2 quickly splits into two $\dot{H}$ radicals at high temperatures. Allyl radical recombination, the process by which allyl radicals in propene mostly combine with H_2 to form allyloxy and hydroxyl radicals, competes to end the chain-terminating process known as allyl radical recombination. Propene does not have an effective mechanism for producing H_2 during oxidation, in contrast to ethylene and hence it is lesser reactive at high temperatures.

In the low-temperature zone, propene is more reactive than ethene due to its greater reliance on the direct oxidation of isopropyl radicals and reactions involving C0–C2 base chemistries, which are the primary mechanisms for producing H_2 radicals. These active sites are more abundant in propene than in ethene, leading to increased reactivity at lower temperatures.

Alcohols like methanol, ethanol, and propanol have shorter ignition delay times than hydrocarbons over a broad temperature range. This increased reactivity is attributed to the hydroxyl functional group (–OH). Among these alcohols, methanol has the shortest ignition delay time, in the order of increasing IDT characteristics by ethanol and propanol respectively, showing a trend of diminishing reactivity as the length of carbon chain increases. Ethanol's reactivity diminishes as the CH_3 radicals formed during the decomposition of $C_2H_5OH \rightleftharpoons \dot{C}H_3 + \dot{C}H_2OH$, undergoes recombination leading to the stable compound C_2H_6. Similarly, in case of propanol, the decomposition of reaction $C_3H_7OH \rightleftharpoons \dot{C}_2H_5 + \dot{C}H_2OH$ produces ethyl radicals, which reacts with OH radicals present in the system ($\dot{C}_2H_5 + \dot{O}H \rightleftharpoons \dot{C}H_3 + CH_3OH$) to form the stable compound C_2H_6 after reconciliation of CH_3 radicals. The observed trends in the characteristics of ignition delay times can be explained by molecular structure, bond strengths, and the presence of different functional groups (Pathak and Panigrahy 2024).

As illustrated in the flux diagram for propanol at compression temperature of 800 K and compression pressure of 50 bar (Fig. 2.8), the hydroxyl group ($\dot{O}$ H) present in alcohols influences the ability of the fuel to react with other reactive species present in the system either through hydrogen abstraction or by some other chemical processes. Propanol is less reactive in comparison to that of propane as it exhibits the formation of aldehydes by following the oxidative reaction pathway of fuel radicals in presence of O_2 (O_2 + $C_3H_6OOH1\text{-}2 \rightleftharpoons 2OH + C_2H_4OCHO$), at lower temperature region, which in a way suppresses the formation of QOOH (O_2 + $C_3H_6OOH1\text{-}2 \rightleftharpoons C_3H_6OOH2\text{-}1O_2$), that is crucial for chain-branching reactions responsible for reactivity in low-temperature region, as illustrated in Fig. 2.9.

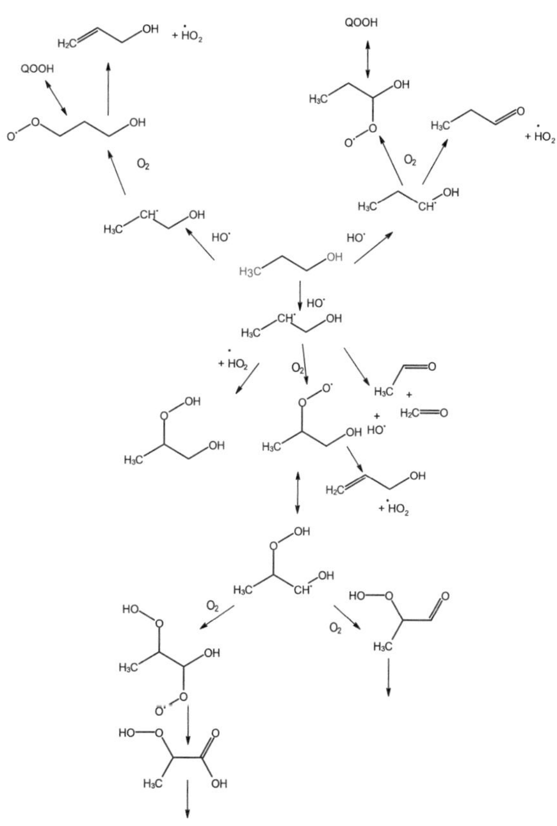

Fig. 2.8 Flux diagram for ignition characteristics of pure n-C_3H_7OH at T_C = 800 K, p_C = 50 bar and 18% fuel consumption (reprinted from Pathak and Panigrahy (2024) with permission of Begell House, Inc.)

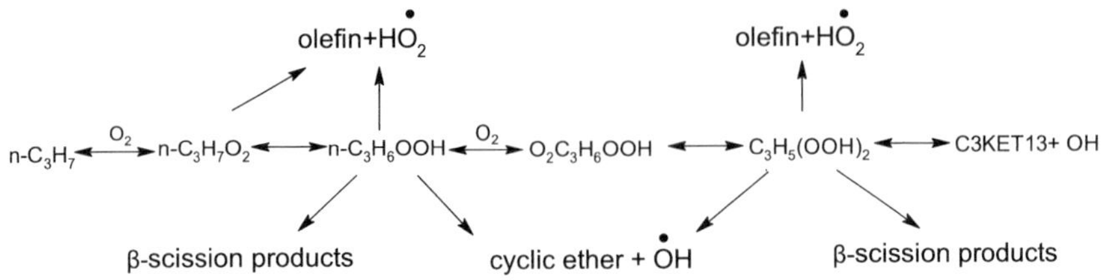

Fig. 2.9 Low-temperature oxidation chemistry of propane (reprinted from Pathak and Panigrahy (2024) with permission of Begell House, Inc.)

Conversely, propane follows a distinct reaction pathway under the same temperature conditions. It undergoes direct oxidation of isopropyl radicals, resulting in the formation of hydrogen (H_2).

A thorough analysis of fundamental properties like IDT characteristics and the associated analysis of how the reactions are proceeding to affect the overall reactivity of the system by acknowledging the pathway which is more dominant over the other provides valuable insights into the varying reactivity of hydrocarbon fuels and alcohols. Comprehending the variability among the reaction mechanism for these different fuels is crucial for optimizing combustion systems and developing fuels or fuel mixture that are more efficient and produce fewer emissions.

2.5 Summary and Conclusions

Governments worldwide have established ambitious frameworks and objectives aimed at attaining net-zero carbon emissions as a means to combat climate change and mitigate its associated risks. These initiatives are under rigorous examination and surveillance by entities such as the United Nations Framework Convention on Climate Change, the International Energy Agency, and the Intergovernmental Panel on Climate Change. In an effort to monitor progress, share insights, and disseminate best practices, various international symposia are convened annually. Notable among these are the United Nations Climate Change Conference, the Global Summit on Climate Action, and the Carbon Pricing Leadership Coalition Annual Conference. Developed countries, including the USA, the European Union, the UK, Japan, and Australia, have pledged to achieve the targets necessary to attain net-zero emissions by 2050, whereas developing nations like China and India have established their objectives for 2060 and 2070, respectively. However, the pathway to achieving net-zero emissions is fraught with formidable challenges. A significant impediment is the substantial cost associated with the deployment of low-carbon or zero-emission technologies. The energy requirements of numerous nations are predominantly satisfied through the utilization of fossil fuels, and transitioning from these sources necessitates extensive investments in novel infrastructure, which could prove prohibitively expensive for countries with limited financial capabilities.

Moreover, the journey toward net-zero is exacerbated by technological hurdles. It demands extensive research, development, and implementation efforts to invent new technologies or adapt existing ones to meet low-carbon or zero-emission standards. These technological difficulties have the capacity to hinder progress and amplify the complexity of transitioning to net-zero. Confronting these challenges necessitates a deep understanding of the reactivity of various fuels, such as alcohols and hydrocarbons. The exploration of ignition delay times and reaction pathways provides invaluable insights into fuel reactivity. For example, it has been identified that ethene exhibits greater reactivity than ethane, primarily due to alkenes containing double bonds, while methane is characterized by the lowest reactivity. Among alcohols, methanol is the most reactive, succeeded by ethanol and propanol. The reactivity of alcohols is affected by the hydroxyl group through mechanisms such as hydrogen abstraction and other chemical reactions. Propane is found to be more reactive than ethane and methane, attributed to its larger molecular structure and greater predisposition to hydrogen abstraction.

The analysis of temperature ranges yields critical insights for practical combustion scenarios, specifically highlighting the superior reactivity of ethanol compared to ethane and ethene within certain thermal conditions. This distinction is pivotal when considering ethanol's efficacy as a biofuel. Understanding the disparity between reactivity and reaction paths is crucial in enhancing combustion processes. This perspective is beneficial for predicting and refining the combustion behavior of C1–C3 compounds across varying temperatures. In summation, the realization of net-zero ambitions is contingent upon surmounting the financial, scientific, and technical challenges. A profound comprehension of fuel reactivity and combustion dynamics plays a key role in achieving these global environmental aims.

Nomenclature

$\dot{R}$	Alkyl
$R\dot{O}_2$	Alkyl peroxyl
$\dot{Q}OOH$	Hydroperoxyl alkyl
$\dot{O}_2QOOH$	Hydroperoxyl alkyl peroxyl
HPST	High-pressure shock tube
NTC	Negative temperature coefficient

References

Ajulu CE (2020) Automated generation of chemical kinetic mechanisms of Fischer-Tropsch (FT) Gas-to-liquid fuel surrogates. Master's thesis, Northeastern University

Allen JW, Goldsmith CF, Green WH (2012) Automatic estimation of pressure-dependent rate coefficients. Phys Chem Chem Phys 14(3):1131–1155

ANSYS CHEMKIN-PRO. https://www.ansys.com/products/fluids/ansys-chemkin-pro
Chehrmonavari H, Kakaee A, Hosseini SE, Desideri U, Tsatsaronis G, Floerchinger G, Paykani A (2023) Hybridizing solid oxide fuel cells with internal combustion engines for power and propulsion systems: a review. Renew Sustain Energy Rev 171:112982
Cuoci A, Frassoldati A, Faravelli T, Ranzi E (2013) Numerical modeling of laminar flames with detailed kinetics based on the operator-splitting method. Energy Fuels 27(12):7730–7753
Cuoci A, Frassoldati A, Faravelli T, Ranzi E (2015) OpenSMOKE++: an object-oriented framework for the numerical modeling of reactive systems with detailed kinetic mechanisms. Comput Phys Commun 192:237–264
Curran HJ (2019) Developing detailed chemical kinetic mechanisms for fuel combustion. Proc Combust Inst 37(1):57–81
Deutschmann O, Tischer S, Correa C, Chatterjee D, Kleditzsch S, Janardhanan VM, Daymo E (2004) DETCHEM software package
Frost J, Hellier P, Ladommatos N (2023) Bio-derived lactones–combustion and exhaust emissions of a new class of renewable fuels. Energy Convers Manage 283:116886
Gao CW, Allen JW, Green WH, West RH (2016) Reaction Mechanism Generator: automatic construction of chemical kinetic mechanisms. Comput Phys Commun 203:212–225
Goodwin DG, Speth RL, Moffat HK, Weber BW (2018) Cantera: an object-oriented software toolkit for chemical kinetics, thermodynamics, and transport processes. Zenodo
Huey C, Metghalchi H, Levendis Y (2024) E-fuels as reduced carbon emission options. ASME Open J Eng 3
IEA (2021) Global CO_2 emissions from transport by subsector, 2000–2030. IEA, Paris. https://www.iea.org/data-and-statistics/charts/global-co2-emissions-from-transport-by-subsector-2000-2030. Licence: CC BY 4.0
IEA (2024) CO_2 Emissions in 2023. IEA, Paris. https://www.iea.org/reports/co2-emissions-in-2023. Licence: CC BY 4.0
Keck JC, Gillespie D (1971) Rate-controlled partial-equilibrium method for treating reacting gas mixtures. Combust Flame 17(2):237–241
Kintechlab:ChemicalWorkBench. http://www.kintechlab.com/products/chemical-workbench/
LOGESOFT. https://logesoft.com
Computational Modelling Cambridge Ltd. The kinetics model builder. http://cmclinnovations.com/solutions/products/kinetics/
Lu T, Law CK (2005) A directed relation graph method for mechanism reduction. Proc Combust Inst 30(1):1333–1341
Magoon GR, Green WH (2013) Design and implementation of a next-generation software interface for on-the-fly quantum and force field calculations in automated reaction mechanism generation. Comput Chem Eng 52:35–45
Marques CS, da Silva JR (2020) Reduced reaction mechanisms for ethanol under ultra-lean conditions in internal combustion engines. ACS Omega 6(1):206–216
Martinez S, Baigmohammadi M, Patel V, Panigrahy S, Sahu AB, Nagaraja S, Curran HJ (2021) A comprehensive experimental and modeling study of the ignition delay time characteristics of ternary and quaternary blends of Methane, ethane, ethylene, and propane over a wide range of temperature, pressure, equivalence ratio, and dilution. Combust Flame 234:111626
Massias A, Diamantis D, Mastorakos E, Goussis DA (1999) An algorithm for the construction of global reduced mechanisms with CSP data. Combust Flame 117(4):685–708
Mohamed AAES, Panigrahy S, Sahu AB, Bourque G, Curran H (2022) The effect of the addition of nitrogen oxides on the oxidation of ethane: an experimental and modelling study. Combust Flame 241:112058
Niemeyer KE, Sung CJ, Raju MP (2010) Skeletal mechanism generation for surrogate fuels using directed relation graph with error propagation and sensitivity analysis. Combust Flame 157(9):1760–1770
Pathak V, Panigrahy S (2024) A comparison of auto-ignition characteristics of alkanes, alkenes, and alcohols ranging from C1 to C3. In: Proceedings of the 27th national and 5th international

ISHMT-ASTFE heat and mass transfer conference, IIT Patna, Patna-801106, Bihar, India. Begel House Inc.
Pepiot-Desjardins P, Pitsch H (2008) An efficient error-propagation-based reduction method for large chemical kinetic mechanisms. Combust Flame 154(1–2):67–81
Pitsch H (1990) FlameMaster v3.3.10: A C++ computer program for 0D combustion and 1D laminar flame calculations. http://www.itv.rwth-aachen.de/index.php?id=128
Rabitz H, Kramer M, Dacol D (1983) Sensitivity analysis in chemical kinetics. Annu Rev Phys Chem 34(1):419–461
Ram V, Salkuti SR (2023) An overview of major synthetic fuels. Energies 16(6):2834
Ramirez A, Sarathy SM, Gascon J (2020) CO_2 derived E-fuels: research trends, misconceptions, and future directions. Trends Chem 2(9):785–795
Razus D, Giurcan V, Movileanu C, Mitu M (2022) Nitric oxide generation in N2-Diluted H2–N2O flames: a computational study. Processes 10(5):1032
Sankaran R, Hawkes ER, Chen JH, Lu T, Law CK (2007) Structure of a spatially developing turbulent lean methane–air Bunsen flame. Proc Combust Inst 31(1):1291–1298
Shi K, Guan B, Zhuang Z, Chen J, Chen Y, Ma Z, Huang Z (2024) Perspectives and outlook of E-fuels: production, cost effectiveness, and applications. Energy Fuels 38(9):7665–7692
Stagni A, Cuoci A, Frassoldati A, Faravelli T, Ranzi E (2014) Lumping and reduction of detailed kinetic schemes: an effective coupling. Ind Eng Chem Res 53(22):9004–9016
Stephen R (2000) Turns. An introduction to combustion: concepts and applications. In: Mechanical engineering series. McGraw Hill, p. 51
Strelkova MI, Kirillov IA, Potapkin BV, Safonov AA, Sukhanov LP, Umanskiy SY, Tentner AM (2008) Detailed and reduced mechanisms of jet a combustion at high temperatures. Combust Sci Technol 180(10–11):1788–1802
Turányi T (1990) Sensitivity analysis of complex kinetic systems. Tools and applications. J Math Chem 5(3):203–248
Wang QD, Panigrahy S, Yang S, Martinez S, Liang J, Curran HJ (2021) Development of multipurpose skeletal core combustion chemical kinetic mechanisms. Energy Fuels 35(8):6921–6927
Warth V, Battin-Leclerc F, Fournet R, Glaude PA, Côme GM, Scacchi G (2000) Computer based generation of reaction mechanisms for gas-phase oxidation. Comput Chem 24(5):541–560
Zheng XL, Lu TF, Law CK (2007) Experimental counterflow ignition temperatures and reaction mechanisms of 1,3-butadiene. Proc Combust Inst 31(1):367–375

Chapter 3
Different Hydrogen Production Methods and the Associated Potential Anthropogenic Hydrogen Releases Along the Entire Hydrogen Value Chain

Nidhi Sharma and Sanjay Singh

Abstract The present chapter is a review on the environmental impact of different anthropogenic hydrogen releases along the entire hydrogen value chain. Hydrogen is becoming one of the most demanding sources of energy as its utilization is emission-free. However, hydrogen handling during various stages of hydrogen value chain involves safety measures due to its unique properties (highly flammable) and potential hazards like harmful emissions post-combustion. The entire hydrogen value chain, starting from production and followed by storage, transportation and utilization involves two major types of emission sources, namely, stack and fugitive emissions. Stack emissions are the flue gases that are released during the production of hydrogen and fugitive emissions are unintentional hydrogen leaks during storage, transportation, and delivery. Several extreme measures have been taken to reduce stack emissions due to their high volumes. Technologies such as CO_2 capture and storage (CCS), scrubbing and adsorption and membrane methods are commonly employed to mitigate CO_2 emissions from industrial processes. Unlike stack, unintentional or fugitive emissions are unpredictable, usually originate from piping equipments and fittings, and are not restricted to a specific source-point. They generally appear as leaks from the containers carrying the process fluid such as through valves, pumps, and flanges. Basically, fugitive emissions are controlled through technical and management approaches in industrial settings. From hydrogen production standpoint, both stack and fugitive emissions needs careful consideration. In order to ensure sustainable and environmentally viable hydrogen production, future advancements and regulatory frameworks should aim to minimize both types of emissions.

Keywords Hydrogen (H_2) · Hydrogen economy · Hydrogen production · Hydrogen storage · Hydrogen transportation · Methane (CH_4) · Carbon dioxide (CO_2)

N. Sharma (✉) · S. Singh
Amity University Uttar Pradesh, Noida, Uttar Pradesh, India
e-mail: nsharma18@amity.edu

N. Sharma et al. (eds.), *Sustainable, Scalable and Storable E-Fuels for Decarbonising Transport Sector*, Energy, Environment, and Sustainability,
https://doi.org/10.1007/978-981-96-7388-9_3

3.1 Introduction

In 2022, among different gases accounting for US greenhouse gas emissions from human activities, carbon dioxide (CO_2), methane (CH_4), and nitrous oxide (N_2O) accounted for 80, 12, and 6%, respectively, making CO_2 as the primary greenhouse gas (GHG) and methane as second most important GHG emitted through human activities (https://www.epa.gov/ghgemissions/overview-greenhouse-gases). Human activities that release methane include leaks from natural gas systems and livestock farming. Methane (CH_4) has a much shorter lifespan in the atmosphere compared to carbon dioxide (CO_2), but it is more effective at trapping radiation than CO_2. Globally, human activities account for 50–65% of total methane (CH_4) emissions. Energy production, industrial operations, waste management, agriculture, and land use all emit methane. After agricultural sources, natural gas and petroleum systems are the second-largest source of CH_4 emissions (https://www.epa.gov/ghgemissions/overview-greenhouse-gases). When natural gas is produced, processed, stored, transmitted, distributed, and used, as well as when crude oil is produced, refined, transported, and stored, methane is released into the atmosphere. Throughout the natural gas systems, there are several individual components that are susceptible to leaks such as compressors, valves, pumps, flanges, gauges, and pipe connectors, among others. Besides unintentional leaks, several sources deliberately vent gas. For instance, pneumatic valves, used across natural gas systems, which function on pressurized natural gas and intentionally release small amounts of gas by design as part of normal operation. In addition, methane gas is often produced from wells following hydraulic fracturing or when liquids are removed, storage tanks, dehydrators, and depressurizing equipment before maintenance.

The burning of fossil fuels (oil, coal, and natural gas) for energy and transportation is the primary human activity that releases CO_2. The International Energy Agency's (IEA) latest analysis (IEA 2024) highlights that global energy-related CO_2 emissions from energy combustion and industrial processes increased by 0.9% in 2022, amounting to an additional 321 million tonnes (Mt) of CO_2, resulting in a new record high of over 36.8 gigatonnes (Gt) of CO_2 emissions. Global Energy Outlook (GEO 2023) (Richter et al. 2023) report stated that limiting global warming to 1.5 °C requires a 37 Gt reduction in CO_2 emissions from 2022 levels and achieving net-zero emissions in the energy sector by 2050. According to IEA report (World Nuclear Association 2024), burning of fossil fuels for generation of electricity is responsible for over 40% of CO_2 emissions. To reduce CO_2 emissions from electricity generation, hydrogen has attracted lot of attention as a potential solution since hydrogen combustion releases only water as a by-product. Thus, hydrogen is seen as a versatile energy carrier, capable of decarbonizing various sectors, including industry, transportation, and power generation. Several countries have formulated policies and initiatives to promote hydrogen production and use (Ma 2023; Oliveira et al. 2021; Atilhan et al. 2021), recognizing its potential to help meet climate targets and achieve energy security.

The global interest in hydrogen has been steadily increasing due to its potential to play a crucial role in the transition to a sustainable, low-carbon energy future. The transition towards a hydrogen-based economy necessitates a comprehensive understanding of the safety implications associated with hydrogen handling across the entire value chain. In recent years, hydrogen is appearing as a possible energy source replacing fossil fuels. While hydrogen is emission-free in its end-use, energy is required for its production which can result into emissions of various types, including fugitive emissions from piping equipment and fittings.

A range of domestic and commercial utilizations are supported through hydrogen, as an essential energy carrier by providing electricity and heating (Parra et al. 2014). Hydrogen powers technologies like hydrogen fuel cells (Abderezzak et al. 2014) are shown in Fig. 3.1, which is used in vehicles (Parra et al. 2014) and other appliances. It can be obtained from renewable sources or fossil fuels (Abderezzak et al. 2014; Cipriani et al. 2014; Forsberg 2009). About 96% of global hydrogen production is currently being obtained from fossil fuels like natural gas and coal (Dufour et al. 2009). The predominant method used on a large scale is steam methane reforming (SMR) or also called as natural gas steam reforming method. In this method, methane is heated along with steam to produce hydrogen and carbon dioxide is generated as a by-product (Nieva et al. 2014). Currently, about 80–85% of the world's hydrogen largely comes from the steam methane reforming (SMR) method (Simpson and Lutz 2007). The rest of the hydrogen production is obtained from other methods such as coal gasification and water electrolysis (Simpson and Lutz 2007). Among all the different methods for hydrogen production, water electrolysis is the cleanest method as it only generates oxygen as a by-product. However, it currently faces economic challenges in competing with established technologies like SMR (Boyano et al. 2011; Edwards et al. 2008). While water electrolysis has clear environmental advantages, it is currently more expensive than SMR due to the higher costs of the electrolysis process and the electricity required. Additionally, the technology is not yet as scalable or commercially widespread, making it less cost-competitive at present, whereas SMR remains cost-effective because it uses an established and widely available infrastructure (natural gas pipelines, reactors, etc.), making it cheaper and more efficient than electrolysis in most cases.

Though hydrogen is a colourless gas, its different types are associated with specific colour designation to indicate and to differentiate between different technologies used to produce each type of hydrogen. The different colours of hydrogen are used to distinguish between the production methods, from conventional fossil fuel-based methods to more sustainable, renewable methods. These colours are determined by the molecule being broken down, the energy source used to extract hydrogen, and the by-products of the chemical reactions. Given the need to reduce greenhouse gas emissions, the discussion around hydrogen colours primarily focuses on the emissions of carbon dioxide and methane. The following Table 3.1 summarizes the different types of hydrogen production methods and the designated colour. It must be noted that these different colours only point towards the production methods, not the hydrogen produced.

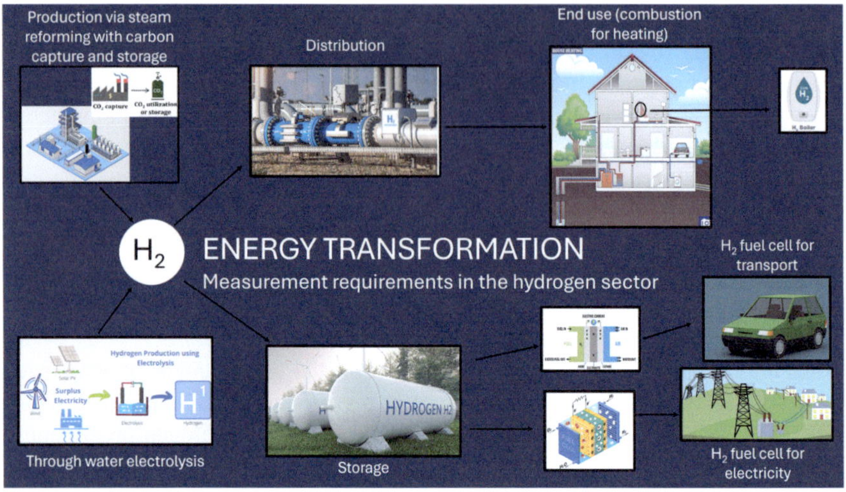

Fig. 3.1 Energy transformation

Table 3.1 Different type of hydrogen production methods and associated colour codes

Green hydrogen	Blue hydrogen	Grey hydrogen
Produced using clean electricity from renewable energy sources like solar and wind, and it emits no damaging greenhouse gases. Yellow hydrogen is another name for hydrogen produced by sun energy	Produced by a process known as steam reforming from natural gas. Carbon dioxide and hydrogen are the by-products of the process. It is crucial to capture and store carbon. Low-carbon hydrogen is another name for it	Most used method of producing hydrogen. Approximately, 96% is made from methane or natural gas via steam methane reformation; however, the greenhouse gas emissions from the process are not captured
Black and brown hydrogen	**Pink hydrogen**	**White hydrogen**
Made from either brown or black coal. They are the most environmentally damaging and the complete opposite of green hydrogen in the hydrogen spectrum	Produced using nuclear energy-driven electrolysis. Red hydrogen or purple hydrogen is other names for nuclear-produced hydrogen	Naturally occurring geological hydrogen produced by hydraulic fracturing and found in subterranean deposits. There are currently no plans to take advantage of the hydrogen

From an environmental standpoint, hydrogen is viewed as a highly promising solution for reducing emissions (Díaz Alvarado and Gracia 2010). However, the production of hydrogen requires energy and obtaining hydrogen from fossil fuels as primary energy source (such as natural gas or coal) can lead to emissions (greenhouse gases). Whether from stack gases or fugitive sources, these emissions pose significant challenges related to the use of fossil fuels for hydrogen production. Consequently, this aspect indirectly suggests that hydrogen may not be an environmentally friendly

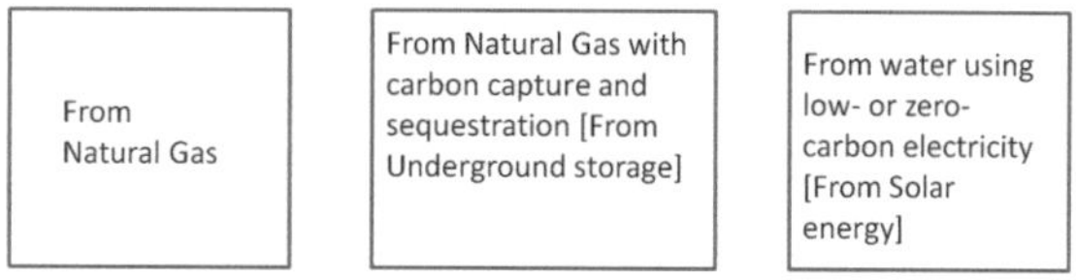

Fig. 3.2 Three major widely used hydrogen production methods (theconversation.com)

energy source. As a result, the production of hydrogen through SMR process and its environmental impact underlines an important disadvantage of the method and raise concerns about its alignment with the environmental regulations. Figure 3.2 shows the three major widely used hydrogen production methods.

In recent times, significant actions from academia and industry have focused on assessing various aspects of hydrogen production, including economic considerations (Cruz and Oliveira 2008), as well as exergy and energy balances (Cruz and Oliveira 2008; Rosen 1995; Rosen and Scott 1992; Tugnoli et al. 2008). More recently, attention has also turned towards understanding the environmental impacts associated with hydrogen production (Granovskii et al. 2006a, b; Kothari et al. 2008; Spath and Mann 2000). These environmental impacts have been quantified and analysed by carrying out various number of studies (Spath and Mann 2000; Koroneos et al. 2004; Wang et al. 2007) and comparing different methods of hydrogen production for instance biomass, solar electricity, water electrolysis, fossil fuels, wind, and hydropower, in terms of their advantages and disadvantages (Granovskii et al. 2006a, b). However, one critical aspect that has often been overlooked in these assessments is fugitive emissions. These emissions, if not properly accounted for, can lead to significant issues such as financial losses, health impacts, and environmental damage over time. Fugitive emissions can arise from various stages of hydrogen production, storage, and transportation and their neglect in environmental assessments may underestimate the true environmental footprint and potential risks associated with hydrogen technologies. Moving forward, it is crucial for future studies and assessments to include comprehensive evaluations of fugitive emissions. This will ensure a more accurate understanding of the environmental impacts of hydrogen production and enable better-informed decision-making towards sustainable hydrogen technologies. This chapter aims to study different hydrogen production technologies, key issues in handling hydrogen along the entire hydrogen value chain, and categorize different emissions and the solutions to control those emissions.

3.1.1 Key Issues in Handling Hydrogen

Handling hydrogen across its entire value chain involves several safety implications due to its unique properties and potential hazards such as:

- **Flammability**: Hydrogen is highly flammable over a wide range of concentrations in air (4–75% by volume). This presents a significant fire and explosion hazard if not handled properly.
- **Leaks**: Hydrogen molecules are very small, which can lead to leaks even through small openings or imperfections in equipment. This increases the risk of fire or explosion, especially in confined spaces.
- **Pressure hazards**: Hydrogen is typically stored and transported under high pressure (up to 700 bars for fueling stations). This requires robust infrastructure and equipment to handle safely.
- **Material compatibility**: Hydrogen can embrittle metals, making them susceptible to cracking and failure over time. Specialized materials such as high-strength alloys or composites are often required.
- **Ignition sources**: Hydrogen can ignite with static electricity, sparks, or even sunlight under certain conditions. Eliminating potential ignition sources is critical in hydrogen handling facilities.
- **Cryogenic hazards**: Liquid hydrogen is extremely cold (−253 °C). Contact with skin or materials not designed for cryogenic temperatures can cause severe frostbite or structural damage.
- **Toxicity concerns**: While hydrogen itself is not toxic, combustion products such as nitrogen oxides (NOx) can be produced during combustion in air, posing environmental and health risks.
- **Hydrogen embrittlement**: Hydrogen can permeate metals, potentially causing embrittlement and structural integrity issues in storage tanks, pipelines, and other infrastructure.
- **Regulatory compliance**: Hydrogen facilities must comply with various safety standards and regulations to ensure safe handling, storage, and transport.

Addressing these safety implications requires rigorous risk assessments, specialized training for personnel, robust safety protocols, and advanced technologies for leak detection, ventilation, and emergency response. The safe deployment of hydrogen across its value chain can be ensured by mitigating the involved risks through continuous research and development which is now feasible due to development of hydrogen technologies.

Another major concern with hydrogen handling is the fugitive emissions, for instance, hydrogen sulphide (H2S). Trace amounts of hydrogen sulphide may be present in hydrogen emissions from certain sources (e.g. biogas reforming), which is toxic and can corrode equipment.

3.1.2 Types of Hydrogen Emissions

Emissions in chemical processes can be categorized into two; via stack (intentional) or fugitive (unintentional) emissions. Both are intrinsic to any industrial method.

Intentional or stack emissions are the flue gases such as CO_2, Nox, SO_2, CO, particulate matter (PM), volatile organic compounds (VOCs), and heavy metals. These are produced during combustion of fuels like coal, wood, fuel oil, or natural gas in industrial furnaces, boilers, steam generators, or large combustion devices, which are then released into the atmosphere (http://en.citizendium.org/wiki/Flue_gas_stack). On the other hand, fugitive emissions are unintended leaks of chemicals, in various physical forms, from industrial sites (Onat 2006; Uk 2009).

Although fugitive emissions are typically not as significant as stack emissions from traditional fuel combustion, they still present environmental and health risks, especially for workers within the plant area. Fugitive hydrogen emissions can contribute to local air quality issues, even though hydrogen itself is non-toxic. The concerns arise primarily from the indirect effects of hydrogen leakage, particularly through the co-released gases and the broader environmental impact. For instance, if hydrogen is produced via natural gas reforming or transported through pipelines alongside methane, the methane emissions could be significant. Methane is a potent greenhouse gas and contributes to the formation of ground-level ozone, which can worsen air quality. Additionally, in industrial or transportation settings, hydrogen leakage might occur alongside volatile organic compounds (VOCs), which can contribute to smog and particulate matter, affecting local air quality and health. Furthermore, hydrogen combustion or use in certain engines might also release nitrogen oxides, which can lead to the formation of ozone and fine particulate matter (PM2.5), both of which are harmful to respiratory health.

Hydrogen itself, when released into the atmosphere, can have a role in atmospheric chemistry, potentially affecting local air quality indirectly. When hydrogen mixes with the atmosphere, it can influence the formation of ozone through reactions with hydroxyl radicals (OH). This process may increase local ozone levels, especially in regions with high sunlight and pollution. Ozone at ground level is a key component of smog and has negative health impacts, particularly on respiratory systems and can worsen conditions for people with asthma or bronchitis. Hydrogen can interact with OH radicals, potentially altering their concentrations in the atmosphere. This can, in turn, affect the breakdown of other pollutants or influence secondary pollutant formation, impacting air quality. Therefore, though hydrogen is non-toxic, its leakage can contribute to air quality issues through co-released gases like methane, VOCs, and NOx, which directly impact local pollution levels. Additionally, hydrogen's role in altering atmospheric chemistry can influence ozone and particulate formation, indirectly worsening air quality. These factors highlight the need for careful monitoring of fugitive hydrogen emissions to minimize environmental and health impacts, especially in areas with heavy industrial or transportation activity.

The major points regarding the emissions associated with hydrogen production and their potential impacts are:

- **Energy Intensity**: Producing hydrogen often involves energy-intensive processes such as steam methane reforming (SMR), electrolysis, or biomass gasification. The source of this energy (e.g. fossil fuels, renewable sources) determines the overall environmental footprint of hydrogen production.

- **Process Emissions**: During hydrogen production, various process emissions can occur, including leaks from piping equipment and fittings. These fugitive emissions may contain hydrogen itself or other gases depending on the production method used.
- **Environmental Impact**: While fugitive emissions from hydrogen production may not be as large in volume as stack emissions, they can still contribute to local air quality issues and, if not properly managed, may have broader environmental implications.
- **Health Risks**: Workers in hydrogen production facilities are particularly vulnerable to exposure to fugitive emissions. Hydrogen is non-toxic, but other gases associated with production processes (e.g. methane, carbon monoxide) can pose health risks if concentrations are high enough.
- **Safety Measures**: Mitigating fugitive emissions requires rigorous monitoring and maintenance of equipment to prevent leaks. Advanced leak detection technologies and regular inspections are crucial to minimize environmental impact and protect worker safety.
- **Regulatory Compliance**: Compliance with environmental regulations and standards is essential for hydrogen production facilities to ensure emissions are within acceptable limits and do not pose undue risks to human health or the environment.

As hydrogen continues to be developed as a viable energy alternative, continued research and technological advancements are crucial to reduce the energy intensity of production processes and minimize emissions. This approach will help maximize the environmental benefits of hydrogen while ensuring safe operations for workers and surrounding communities.

3.1.3 Methods to Minimize Stack Emissions

Efforts to control stack emissions have been extensive due to their significant volume. Technologies such as CO_2 capture and storage (CCS) are commonly employed to mitigate CO_2 emissions from industrial processes (Dufour et al. 2009). Another developed technology for controlling stack or unintentional emissions is scrubbing, where absorbents in the form of sprays are injected through scrubbers to capture pollutants such as sulphur dioxide (SO_2) and nitrogen oxides (NOx) to capture the impurities (Chen et al. 2012).

Adsorption and membrane technologies are both pivotal in the realm of carbon dioxide (CO_2) removal from various industrial processes, particularly in the context of hydrogen production.

3.1.4 Adsorption Technology

Pre-combustion CO_2 removal: Adsorption technologies such as Electrical Swing Adsorption (ESA), Temperature Swing Adsorption (TSA), and Pressure Swing Adsorption (PSA) are utilized to capture CO_2 before it undergoes combustion. These methods involve passing the gas stream through adsorbent materials that selectively trap CO_2 while allowing other gases to pass through. PSA cycles the pressure, TSA varies the temperature, and ESA applies an electric field to enhance CO_2 capture efficiency.

3.1.5 Membrane Technology

Post-combustion CO_2 removal: After combustion processes, selective membranes are employed to segregate CO_2 from flue gases (Mondal et al. 2012). These membranes are designed to allow CO_2 to selectively permeate through while blocking other gases.

Pre-combustion CO_2 removal: Before combustion occurs, membranes can be utilized to separate CO_2 from hydrogen streams. This ensures that the hydrogen produced is cleaner and suitable for various applications, including fuel cells.

Both adsorption and membrane technologies offer effective solutions for CO_2 removal at different stages of industrial processes. They play a major role in reducing greenhouse gas emissions and improving the environmental sustainability of hydrogen production and other industries reliant on fossil fuels. These technologies continue to evolve, aiming for higher efficiency, lower costs, and broader applicability in mitigating climate change impacts.

3.1.6 Methods to Reduce Fugitive Emissions

While stack emissions are often targeted for reduction through these technologies, fugitive emissions pose a different challenge. They can occur unpredictably from equipment leaks, valves, or other points in industrial facilities, potentially leading to environmental and safety concerns if not properly managed. The same technologies that are used for controlling the stack emissions cannot be used for capturing the fugitive emissions. Unlike stack, unintentional or fugitive emissions are unpredictable, usually originate from piping equipments and fittings and are not restricted to a specific source-point. They generally appear as leaks from the containers carrying the process fluid such as through valves, pumps, and flanges (Uk 2009), as shown in Fig. 3.3. Other cases where fugitive emissions can occur are from unsealed sources such as storage tanks, open-ended lines, pressure relief valves, or vents (Uk 2009).

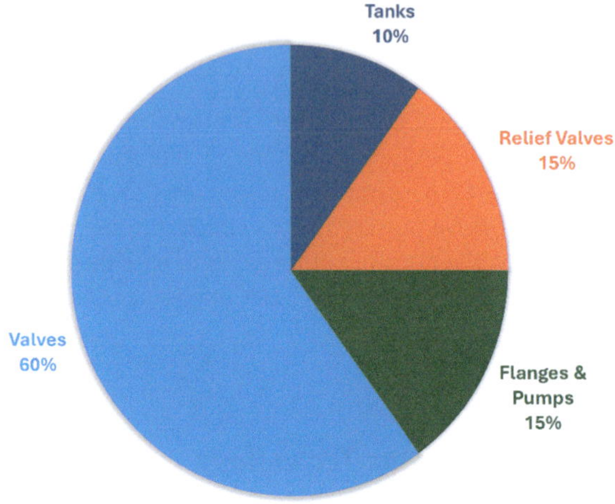

Fig. 3.3 Industrial fugitive emissions sources

From around the world, industrial activities emit approximately more than a million metric tonnes of chemical materials per year as fugitive emissions (Hassim et al. 2010).

It was reported that the volatile organic components (VOCs) fugitive emissions were about 17% of the total VOCs emissions in the year 1996 in Europe (Uk 2009). Since the fugitive emissions not only impact the environment and worker's health conditions but also causes financial losses as the released materials are of potential value (Uk 2009; Hassim et al. 2010), it becomes significantly important to reduce fugitive emissions. Essentially, controlling fugitive emissions involves both technical and management approaches aimed at reducing leaks and improving overall environmental performance in industrial settings.

3.1.7 Technical Approaches

Advanced Equipment and Technology: Utilizing innovative processes such as high-efficiency distillation and reactive distillation units can minimize emissions. Employing high quality canned pumps and seal-less valves, which have improved sealing mechanisms, helps prevent leaks.

Improvements in Process Operations: Implementing measures such as sealing potential emission points for instance, open-ended lines are sealed using blinds, caps or plugs. Enclosing seal areas or vents and directing emissions to combustion control devices can also mitigate fugitive emissions. Operating processes under vacuum conditions where feasible further reduces the likelihood of leaks (EPA 1986, 1994).

3.1.8 Management Approaches

Leak Detection and Repair (LDAR) Programmes: These programmes are crucial for continuously monitoring and promptly repairing leaks in petrochemical and other industries. Regular inspections and use of sensitive detection technologies help identify leaks early, preventing significant emissions (EPA 1986, 1994).

Regulatory Compliance: Regulations such as Directive 2008/1/EC of the European Parliament and Council focus on integrated pollution prevention and control (IPPC). Industries must apply for permits that detail emission sources, quantities, and proposed techniques for monitoring and reducing emissions. Compliance ensures that industries adhere to standards aimed at minimizing environmental impact (Directive 2008).

Awareness and Knowledge Enhancement: Efforts are also focused on raising awareness and enhancing knowledge about fugitive emissions. Many emissions are overlooked due to a lack of understanding or awareness of their impact. Education and training programmes play a vital role in ensuring that industries prioritize emission reduction and adopt best practices.

By combining technical innovations, effective management practices like LDAR programmes, regulatory frameworks, and increased awareness, industries can significantly reduce fugitive emissions. These efforts are crucial for achieving environmental sustainability and minimizing the environmental and financial impacts associated with industrial operations.

Furthermore, hydrogen leak detection is critical for safety in industries using hydrogen, such as energy, manufacturing, and transportation. The nature of hydrogen being colourless, odourless, and highly flammable makes detecting leaks a challenge. Electrochemical sensors or sensor platforms that use palladium thin films, where hydrogen adsorption modifies the electrical and optical characteristics of the film, make it simple to detect hydrogen (https://explore.mission-innovation.net/mission/clean-hydrogen/). Each traditional hydrogen detection technology has specific optimal operating conditions, along with performance specifications, including the range of hydrogen concentrations, measurement response time, and resistance to chemical interferences. Other new advanced hydrogen detection technologies are fibre optic sensor, ultrasonic leak detectors, acoustic sensor concept, Schlieren imaging, and memory-based sensor.

The fundamental principle behind fibre optic sensors is the detection of changes in light interference. Those changes are due to physical or chemical alterations on the fibre's surface. Sensing elements can be positioned either at specific points along the fibre (example at the tip) or arranged at short intervals along its length. These sensor components can respond chemically to hydrogen or physically to a leak, like tension or audio signals (https://explore.mission-innovation.net/mission/clean-hydrogen/).

Fibre Bragg Grating (FBG) is a technique used in fibre optic sensors that uses a tiny layer of palladium (Pd) or Pd alloy on the fibre surface. This thin layer will swell with hydrogen adsorption, changing the wavelength shift of FBG (Fig. 3.4). It

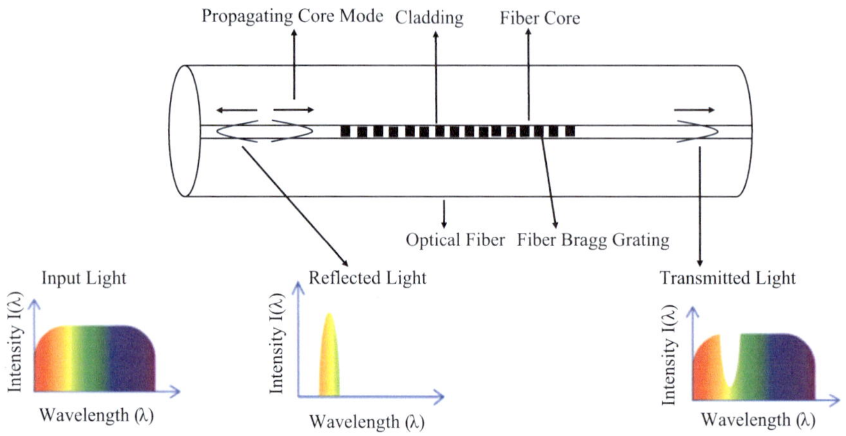

Fig. 3.4 Schematic diagram of an FBG hydrogen sensor (Mihailov 2012)

may also be possible to monitor hydrogen pipelines using a fibre optic sensor with dispersed detecting elements.

Ultrasonic leak detectors are designed to detect acoustic signals generated when pressurized gas escapes through a leak (https://explore.mission-innovation.net/mission/clean-hydrogen/). These devices are optimized to identify ultrasonic leak signatures by distinguishing them from operational and background noises, often based on the duration and intensity of the signal. Leaks produce louder and longer sounds than other operational noises. While commercial systems are available, their use in hydrogen facilities is limited. Current research is focused on using multiple ultrasonic systems to pinpoint the source of leaks. The speed of sound in a low-density medium is the basis for another acoustic sensor concept that is still in the development stage. Ultrasonic sensors can detect changes in sound velocity caused by a rise in hydrogen concentration because hydrogen's sound velocity is approximately four times quicker than air's (Fig. 3.5). Both fibre optic sensors and ultrasonic sensors have no-risk for explosion, high durability, and high corrosion-resistance.

In the context of hydrogen production, both stack emissions (e.g. from fossil fuel-based processes like SMR) and fugitive emissions (e.g. from leaks in pipelines

Fig. 3.5 Schematic diagram of an ultrasonic H2 sensor (Nisshinbo Holdings Inc. 2017)

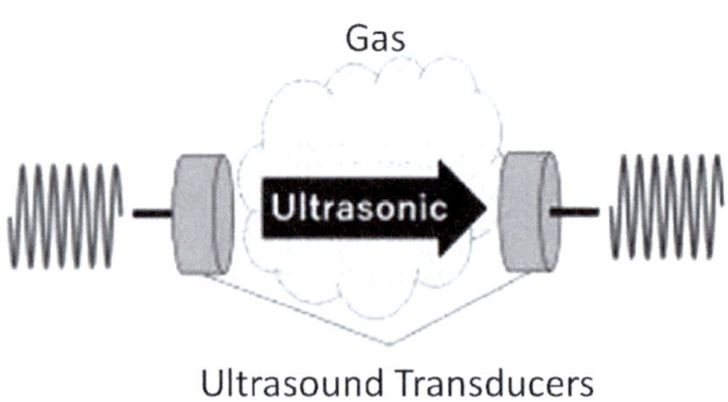

or storage tanks) need careful consideration. Future advancements and regulatory frameworks should aim to minimize both types of emissions to ensure sustainable and environmentally responsible hydrogen production.

3.1.9 *Artificial Intelligence Powered Advancements in Hydrogen Safety*

Conventional hydrogen production methods are dependent on processing of fossil fuels and water electrolysis. Green hydrogen (H_2) is the production of H_2 using water electrolysis powered by renewable energy sources and is considered an ideal alternative to steam reforming of fossil fuels. Hydrogen being a light molecule must be stored under high pressure for safety, and due to high diffusivity and permeability of H_2, the probability of leakage is high. Leak detection, explosion risk reduction, material compatibility evaluation, storage system optimization, infrastructure development planning, dispersion behaviour prediction, combustion characteristic analysis, public perception management, emergency response planning, and environmental impact assessment are the main safety concerns in H2 systems. A recent study by Patil et al. (2024) provides a thorough examination of recent advancements in hydrogen safety, with a particular focus on the use of AI methodologies to assess potential safety scenarios and design safety measures. The paper presents a critical analysis of the application of AI methodologies, such as artificial neural networks (ANN), machine learning algorithms, computer vision (CV), and data fusion techniques, in improving H_2 safety measures. The review highlights the transformative potential of AI in revolutionizing safety frameworks by exploring the integration of wireless sensor networks and AI for real-time monitoring, as well as the use of computer vision (CV) to interpret visual indicators related to hydrogen leakage issues. The report also identifies and addresses the primary constraints, including the optimization of AI models for a variety of environmental circumstances and the absence of standardized datasets, simultaneously recognizes the key areas for continued research and development. With the application of AI methodologies, the review anticipates enhanced safety for H_2 related applications (Patil et al. 2024). Artificial intelligence holds the potential to improve green hydrogen production, storage to delivery, and end-use, i.e. through the entire H_2 value chain. Figure 3.6 shows the key applications of AI for advancing green hydrogen production.

3.1.10 *Environmental Impact of Anthropogenic Hydrogen Releases*

Understanding hydrogen emissions is critically important for hydrogen (H_2) to be used widely (Rogelj et al. 2018). As mentioned earlier, it can be obtained from

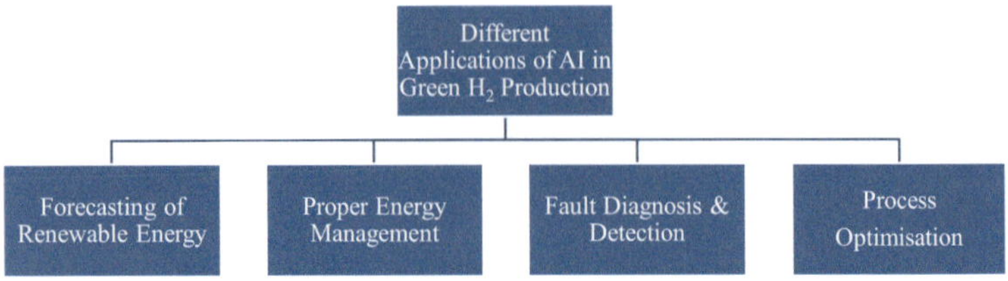

Fig. 3.6 Different applications of AI in green H_2 production

renewable sources or fossil fuels (Abderezzak et al. 2014; Cipriani et al. 2014; Forsberg 2009). About 96% of global hydrogen production is currently being obtained from fossil fuels like natural gas and coal (Dufour et al. 2009), as shown in Fig. 3.6 (Ishaq et al. 2022). The predominant method used on a large scale is steam methane reforming (SMR) or also called as natural gas steam reforming method (Nieva et al. 2014). Muradov and Veziroğlu (2005) reported that one kg of net hydrogen production by SMR process can cause a global warming potential (GWP) of 13.7 kg carbon dioxide equivalent (CO_{2eq}). Using pre-calculated modules-based approach, the fugitive emissions of greenhouse gases (GHGs) like methane (CH_4), carbon dioxide (CO_2), and carbon monoxide (CO) from hydrogen production process through SMR were calculated by Yousef et al. (Muradov and Veziroğlu 2005). The study reported that in the breathable zones, the concentrations of GHGs and CO as fugitive emissions were below the threshold exposure limit value. Hence, the outdoor SMR plant with proper ventilation is considered a safer workplace for workers (Alhamdani et al. 2017), from an occupational health perspective (Fig. 3.7).

Cooper et al. (2022) carried out literature quality evaluation and sensitivity analysis to estimate the possible hydrogen (H_2) emissions from various types of hydrogen

Fig. 3.7 Global energy supply share in 2018 (Rogelj et al. 2018)

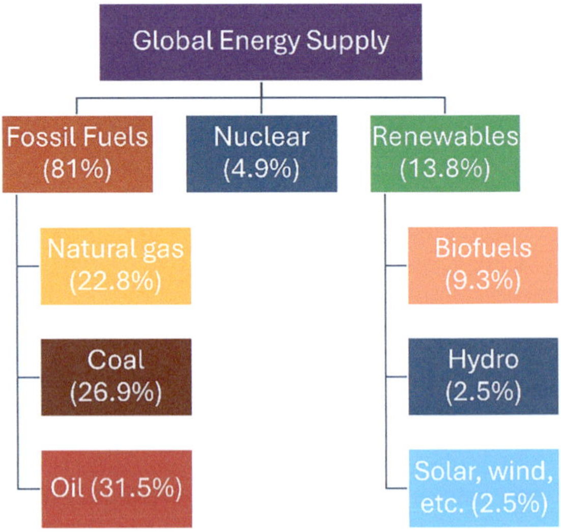

supply chains. The calculations revealed that hydrogen emissions are significantly smaller as compared to that of natural gas if hydrogen is used and operated in the same way as the natural gas due to the lower global warming potential of H_2. Additionally, H_2 has higher energy density implying that for the same end-use, a smaller quantity needs to be transported, resulting in lesser emissions. The study predicted emissions up to 138 million metric tonnes of carbon dioxide equivalent (Mt CO_{2eq}) from H_2 emissions alone in low-carbon energy systems in 2050. Therefore, CO_2 and non-CO_2 emissions are more significant in some hydrogen supply chains. Due to inadequate literature, further work is required to accurately estimate losses and emissions from each stage of the hydrogen supply chain to get a better understanding of its emission profile and its impact to global warming (Alhamdani et al. 2017). Moreover, to address the considerable data gaps, the authors suggested performing experiments to calculate H_2 emissions during productions and processing, loading and unloading and to measure leaks in H_2 pipelines during transmission and distribution.

Cornell and Stanford Universities recently published an article (energym.io) reporting that the production of hydrogen through steam reforming (blue hydrogen) may generate more GHGs emissions as shown in the Fig. 3.5. The study states that the carbon capture and storage (CCS) will not basically halt fugitive emissions of methane and upstream emissions of carbon dioxide from releasing. The study insists that the carbon footprint from the production of hydrogen by steam reforming could be about 20% higher than utilizing natural gas or coal directly for heat or approximately 60% larger than employing diesel oil for heat (Fig. 3.8).

Grey hydrogen, blue hydrogen with carbon dioxide capture from the SMR process but not from the exhaust flue gases produced by burning natural gas to power the SMR equipment, blue hydrogen with carbon dioxide capture from both the SMR process and the exhaust flue gases, natural gas burned for heat generation, diesel oil burned for heat, and coal burned for heat are the sources of the various bars in Fig. 3.5 that represent carbon dioxide equivalent emissions. Emissions of carbon dioxide, including those from the production, processing, and transportation of the fuels, are depicted in orange in the figure. Red indicates fugitive, unburned methane emissions that are equivalent to carbon dioxide. The rate of methane leakage is 3.5% (Howarth and Jacobson 2021).

3.2 Summary and Conclusions

The present chapter provides a review of the different hydrogen (H_2) production methods and anthropogenic releases of H_2 in the environment along the entire H_2 value chain, i.e. from production, storage to delivery, and end-use. While hydrogen offers significant promise as a clean energy carrier, its production involves various emissions and environmental impacts that must be addressed through technological advancements, effective management practices, and regulatory frameworks. For instance, production stage involves CO_2 emissions (also called stack emissions) from

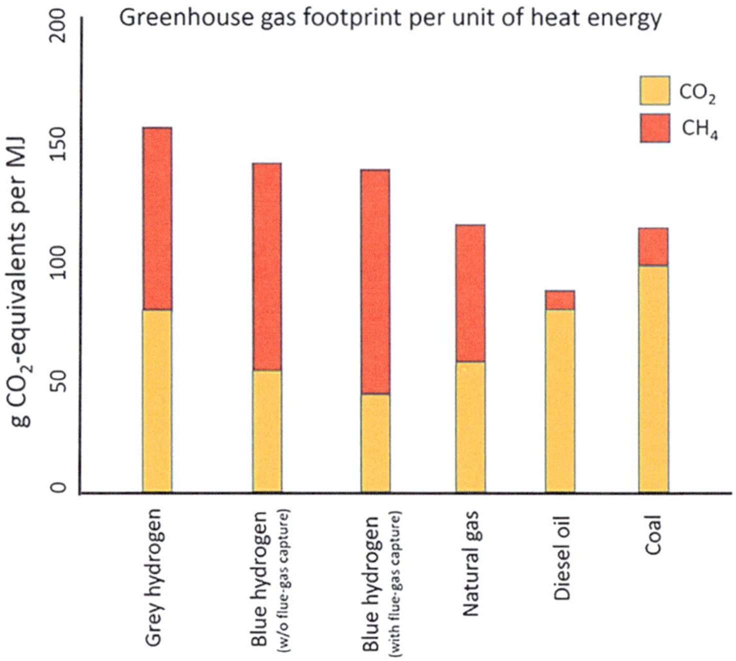

Fig. 3.8 Greenhouse gas footprint per unit of heat energy (Howarth and Jacobson 2021)

fossil fuel-based methods like steam methane reforming (SMR) and coal gasification. Energy use and emissions usually vary based on production method and energy source. After production, storage and transportation stage involves fugitive emissions from pipelines, storage tanks, and equipment leaks, which are challenging to control compared to stack emissions. During the final stage of distribution and utilization, leakage risks persist especially in older infrastructure or improperly maintained systems. It is important to reduce fugitive emissions to maximize the benefits of hydrogen thus, some of the mitigation strategies include technological innovations with implementing advanced equipment (e.g. seal-less valves), and process improvements (e.g. vacuum operations) to minimize fugitive emissions. Additionally, management practices such as adoption of leak detection and repair (LDAR) programmes must be employed and the regulatory compliance, and rigorous monitoring is required to reduce emissions across the value chain. Including environmental considerations such as enhanced awareness and regulatory frameworks to address both direct emissions (CO_2) and fugitive emissions ensures sustainable hydrogen production and use. Ongoing research and development are essential to optimize production methods and mitigate the environmental footprint of hydrogen throughout its entire value chain to achieve a cleaner, safer and more sustainable hydrogen economy. Additionally, the chapter provides an overview of the applications of artificial intelligence (AI)-driven innovations in hydrogen safety. By combining insights from diverse research articles, this chapter acts as a guide for researchers, engineers, and practitioners aiming to advance hydrogen safety technologies.

References

Abderezzak B, Khelidj B, Tahar AM (2014) Modeling charge transfer in a PEM fuel cell using solar hydrogen. Int J Hydrogen Energy 39:1593–1603

Alhamdani YA, Hassim MH, Rex TL (2017) Ng, Hurme M, The estimation of fugitive gas emissions from hydrogen production by natural gas steam reforming. Int J Hydrogen Energy 42(14):9342–9351

Atilhan S et al (2021) Green hydrogen as an alternative fuel for the shipping industry. Curr Opin Chem Eng 31:100668

Boyano A, Blanco-Marigorta AM, Morosuk T, Tsatsaronis G (2011) Exergoenvironmental analysis of a steam methane reforming process for hydrogen production. Energy 36:2202–2214

Carbon dioxide emissions from electricity. World Nuclear Association (2024)

Chen W-H, Hou Y-L, Hung C-I (2012) A study of influence of acoustic excitation on carbon dioxide capture by a droplet. Energy 37:311–321

Cipriani G, Di Dio V, Genduso F, La Cascia D, Liga R, Miceli R, Rico GG (2014) Perspective on hydrogen energy carrier and its automotive applications. Int J Hydrogen Energy 39:8482–8494

CO_2 emissions in 2022—analysis. IEA (2024)

Cooper J, Dubey L, Bakkaloglu S, Hawkes A (2022) Hydrogen emissions from the hydrogen value chain-emissions profile and impact to global warming. Sci Total Environ 830:154624. ISSN 0048-9697

Cruz FE, Oliveira S Jr (2008) Petroleum refinery hydrogen production unit: exergy and production cost evaluation. Int J Thermodyn 11:187–193

Díaz Alvarado F, Gracia F (2010) Steam reforming of ethanol for hydrogen production: thermodynamic analysis including different carbon deposits representation. Chem Eng J 165:649–57.

Directive EU (2008) Directive 2008/1/EC of the European Parliament and of the Council of 15 January 2008 concerning integrated pollution prevention and control. Off J Eur Union L24:8–29

Dufour J, Serrano D, Galvez J, Moreno J, Garcia C (2009) Life cycle assessment of processes for hydrogen production. Environmental feasibility and reduction of greenhouse gases emissions. Int J Hydrogen Energy 34:1370–1376

Edwards PP, Kuznetsov VL, David WIF, Brandon NP (2008) Hydrogen and fuel cells: towards a sustainable energy future. Energy Policy 36:4356–4362

EPA (1986) Emission factors for equipment leaks of VOC and HAP, EPA/450/3-93-002. US Environmental Protection Agency

EPA (1994) Control techniques for fugitive VOC emissions from chemical process facilities, EPA/625/R-93/005. US Environmental Protection Agency

Forsberg CW (2009) Sustainability by combining nuclear, fossil, and renewable energy sources. Prog Nucl Energy 51:192–200

Granovskii M, Dincer I, Rosen MA (2006a) Environmental and economic aspects of hydrogen production and utilization in fuel cell vehicles. J Power Sour 157:411–421

Granovskii M, Dincer I, Rosen MA (2006b) Life cycle assessment of hydrogen fuel cell and gasoline vehicles. Int J Hydrogen Energy 31:337e52

Hassim MH, Perez AL, Hurme M (2010) Estimation of chemical concentration due to fugitive emissions during chemical process design. Process Saf Environ Prot 88:173–184

Howarth RW, Jacobson MZ (2021) How green is blue hydrogen? Energy Sci Eng. 00:1–12. https://doi.org/10.1002/ese3.956

http://en.citizendium.org/wiki/Flue_gas_stack (August 10, 2014)

https://explore.mission-innovation.net/mission/clean-hydrogen/

https://www.epa.gov/ghgemissions/overview-greenhouse-gases
Ishaq H, Dincer I, Crawford C (2022) A review on hydrogen production and utilization: challenges and opportunities. Int J Hydrogen Energy 47:26238–26264
Koroneos C, Dompros A, Roumbas G, Moussiopoulos N (2004) Life cycle assessment of hydrogen fuel production processes. Int J Hydrogen Energy 29:1443–1450
Kothari R, Buddhi D, Sawhney RL (2008) Comparison of environmental and economic aspects of various hydrogen production methods. Renew Sustain Energy Rev 12:553–563
Ma N et al (2023) Large scale of green hydrogen storage: opportunities and challenges. Int J Hydrogen Energy
Mihailov SJ (2012) Fiber Bragg grating sensors for harsh environments. Sensors 12(2):1898–1918
Mondal MK, Balsora HK, Varshney P (2012) Progress and trends in CO_2 capture/separation technologies: a review. Energy 46:431–441
Muradov NZ, Veziroğlu TN (2005) From hydrocarbon to hydrogen-carbon to hydrogen economy. Int J Hydrogen Energy 30:225–237
Nisshinbo Holdings Inc. Notice regarding ultrasonic hydrogen gas sensor (November 7, 2017)
Nieva MA, Villaverde MM, Monzon A, Garetto TF, Marchi AJ (2014) Steam-methane reforming at low temperature on nickel-based catalysts. Chem Eng J 235:158–166
Oliveira AM, Beswick RR, Yan Y (2021) A green hydrogen economy for a renewable energy society. Curr Opin Chem Eng 33:100701
Onat A (2006) A review of fugitive emissions. Seal Technol 2006:7–9
Parra D, Gillott M, Walker GS (2014) The role of hydrogen in achieving the decarbonization targets for the UK domestic sector. Int J Hydrogen Energy 39(9):4158–4169. ISSN 0360-3199
Patil RR, Calay RK, Mustafa MY, Thakur S (2024) Artificial intelligence-driven innovations in hydrogen safety. Hydrogen 5:312–326. https://doi.org/10.3390/hydrogen5020018
Richter M et al (2023) Extinction strain rates of premixed ammonia/hydrogen/nitrogen- air counterflow flames. Proc Combust Inst 39(2):2027–2035
Rogelj J, Shindell D, Jiang K, Fifita S, Forster P, Ginzburg V, Handa C, Kheshgi H, Kobayashi S, Kriegler E, Mundaca L, Séférian R, Vilariño MV (2018) Mitigation pathways compatible with 1.5 °C in the context of sustainable development. In: Masson Delmotte V, Zhai P, Pörtner H-O, Roberts D, Skea J, Shukla PR, Pirani A, Moufouma-Okia W, Péan C, Pidcock R, Connors S, Matthews JBR, Chen Y, Zhou X, Gomis MI, Lonnoy E, Maycock T, Tignor M, Waterfield T (eds) Global warming of 1.5 °C. An IPCC special report on the impacts of global warming of 1.5 °C above pre-industrial levels and related global greenhouse gas emission pathways, in the context of strengthening the global response to the threat of climate change, sustainable development, and efforts to eradicate poverty. IPCC. https://publications.pikpotsdam.de/pubman/faces/ViewItemFullPage.jsp?itemId=item_22900
Rosen MA (1995) Energy and exergy analyses of electrolytic hydrogen production. Int J Hydrogen Energy 20:547–553
Rosen MA, Scott DS (1992) Exergy analysis of hydrogen production from heat and water by electrolysis. Int J Hydrogen Energy 17:199–204
Simpson AP, Lutz AE (2007) Exergy analysis of hydrogen production via steam methane reforming. Int J Hydrogen Energy 32:4811–4820
Spath PL, Mann MK (2000) Life cycle assessment of hydrogen production via natural gas steam reforming. National Renewable Energy Laboratory Golden, CO
theconversation.com
Tugnoli A, Landucci G, Cozzani V (2008) Sustainability assessment of hydrogen production by steam reforming. Int J Hydrogen Energy 33:4345–4357
UK E (2009) Sealing technology-BAT guidance notes, ESA Publication No. 014/09. European Sealing Association, UK
Wang G, Ogden JM, Nicholas MA (2007) Lifecycle impacts of natural gas to hydrogen pathways on urban air quality. Int J Hydrogen Energy 32:2731–2742

Chapter 4
Hydrogen Production and Injection Techniques in IC Engine

Vishnu Agrawal, Sunil, Dilip Sharma, and Nikhil Sharma

Abstract The growing energy demand, fossil fuel depletion, climate change, and stringent emission norms in transportation sector emphasis to think about alternate renewable and clean fuel. Among various alternative fuels hydrogen is quite attractive as it is carbon free and renewable in nature. Hydrogen is an eco-friendly, reliable, and safe energy source found abundant in nature. This chapter examines different methods of producing hydrogen from natural gas reforming, coal gasification, partial oxidation of hydrocarbons, as well as electrolysis from renewable sources like solar, wind, and hydroelectric energy. The production of hydrogen from renewable sources increases flexibility and improves economics with negligible net greenhouse gas emissions. Apart from production, the innovative application of hydrogen as fuel in ICEs is described herein through advanced injection techniques: port fuel injection, direct injection, and dual-fuel systems. Such methods are also associated with improvements in the efficiency of combustion, reductions in emissions, and improved performance in an engine. The novelty also comes from a systematic analysis based on hydrogen's potential as a substitute for traditional fuels with the challenges that lie before, such as the risks of pre-ignition and knock, NOx emissions, and difficulties in storage. Highlighting the transformative potential of hydrogen, this chapter underscores its role in opening the door to cleaner and more sustainable transportation technologies. Future research is encouraged to develop injection systems for hydrogen, thereby overcoming practical barriers in its adoption. This work uniquely combines comprehensive insights into hydrogen production with its advanced application in ICEs, promising a significant leap towards carbon–neutral energy solutions.

Keywords Production of hydrogen · Internal combustion engine · Hydrogen injection techniques

V. Agrawal (✉) · Sunil · D. Sharma · N. Sharma
Department of Mechanical Engineering, Malaviya National Institute of Technology Jaipur, Jaipur, Rajasthan, India
e-mail: 2021rme9557@mnit.ac.in

N. Sharma et al. (eds.), *Sustainable, Scalable and Storable E-Fuels for Decarbonising Transport Sector*, Energy, Environment, and Sustainability,
https://doi.org/10.1007/978-981-96-7388-9_4

4.1 Introduction

Energy demand increased exponentially due to rapidly growing population and development. Presently, most of this demand is fulfilled by traditional fuels, which have played a very important role in boosting the industrial revolution worldwide. However, fossil fuel has some disadvantages, primarily carbon footprint in environment and increased global warming due to CO_2 emission (Ishaq et al. 2022). The prices of fossils fuel are uncertain due to dependency on certain countries. Also, these fossils fuels are depleting very fast and creating the energy in-security.

As shown in Fig. 4.1, the overall transport sector is the leading consumer of oil worldwide. It accounts for about 65% of the total demand for oil. Oil-based fuels including gasoline, diesel, and jet fuel are the key sources of energy that power many modes of transportation such as road vehicles, aviation, shipping, and railways. The sector remains largely dependent on oil, largely because of the dominance of ICE vehicles, which are used as a source of transportation all over the world, even as the adoption of alternative sources of energy is gaining popularity.

The energy demand in the transportation sector is increasing primarily due to the rapid growth in vehicle ownership and global economic development. As urbanization and population growth accelerate, there is a higher demand for personal and commercial transportation to support daily commutes, freight movement, and logistics. Additionally, economic progress in emerging economies has led to greater affordability and accessibility of vehicles, further boosting energy consumption. As shown in Fig. 4.2, the energy demand for passenger vehicles and freight vehicles has increased from 0.9 to 2 EJ and 0.7 to 1.8 EJ respectively in the span of 2005–2020 (Kamboj et al. 2022). The fossil fuels are depleting rapidly. The fossil fuel oil reservoir will deplete by 2050 according to world energy outlook report as shown in Fig. 4.3.

Figure 4.4 shows a significant and steady increase in global CO_2 emissions from the early 1900s–2023. This high acceleration in CO_2 emissions since the mid-twentieth century is associated with extensive use of fossil fuels in energy generation

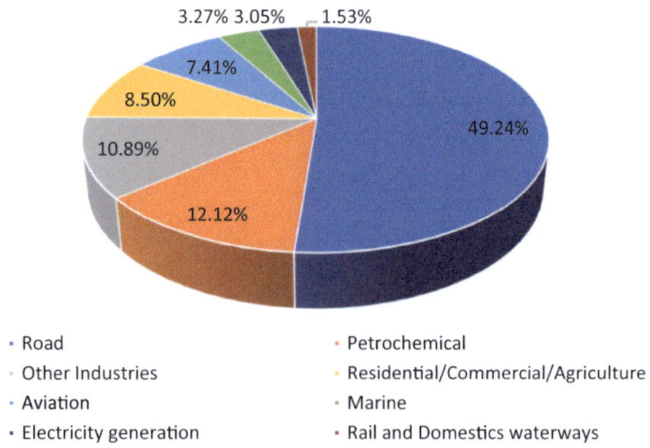

Fig. 4.1 Oil consumption by sector as on 2022 (Tongur 2018)

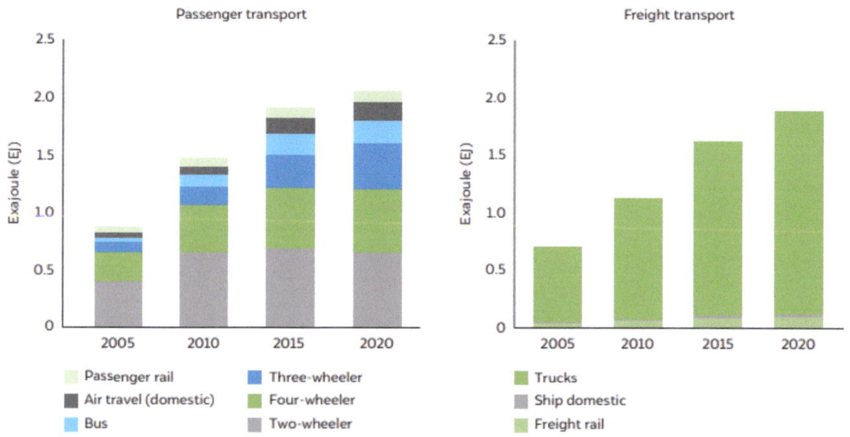

Fig. 4.2 Energy demand in transportation sector from 2005–2020 (Kamboj et al. 2022)

Fig. 4.3 Trend of fossil fuel depletion over time

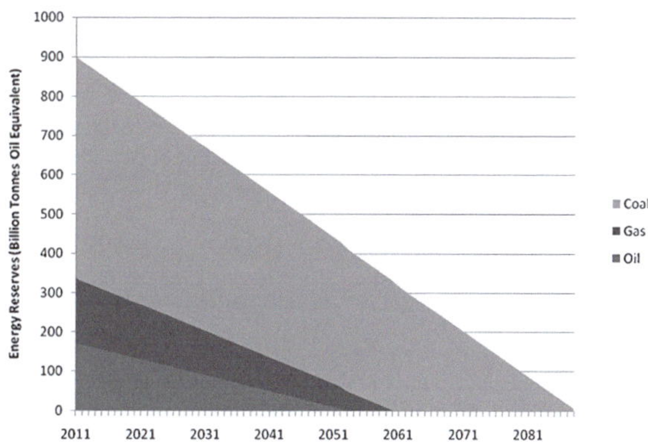

and transportation in the form of coal, oil, and natural gas. This is one of the major causes of climate change due to the fact that carbon dioxide is a major greenhouse gas responsible for trapping heat within the atmospheric system. Rising CO_2 levels have impacts on global warming, extreme weather events, melting ice caps, and sea level rise. Ecosystems are also affected, thus posing significant environmental and socio-economic challenges globally. Among the immediate actions necessary are the transition to renewable energy sources, improvement in energy efficiency, and the development of carbon capture technologies (https://www.iea.org/reports/co2-emissions-in-2023/executive-summary).

The growing energy demand, fossil fuel depletion, climate change, and stringent emission norms in transportation sector emphasis to think about alternate renewable and clean fuel. Among various alternative fuels hydrogen is quite attractive as it is carbon free and renewable in nature. Hydrogen is an eco-friendly, reliable, and safe energy source found abundant in nature. Hydrogen is known as good alternative

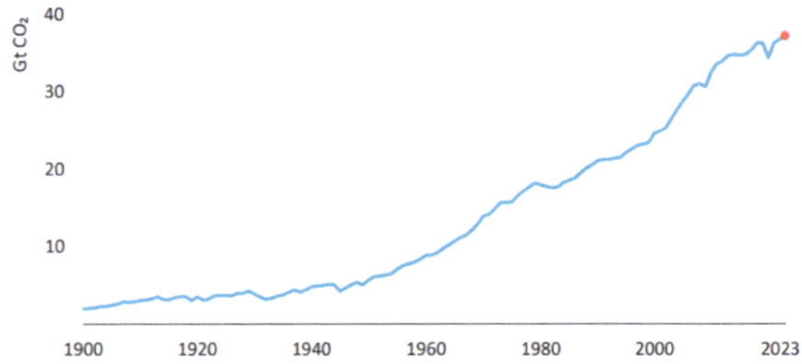

Fig. 4.4 Global energy-related CO_2 emissions and their annual change, 1900–2023

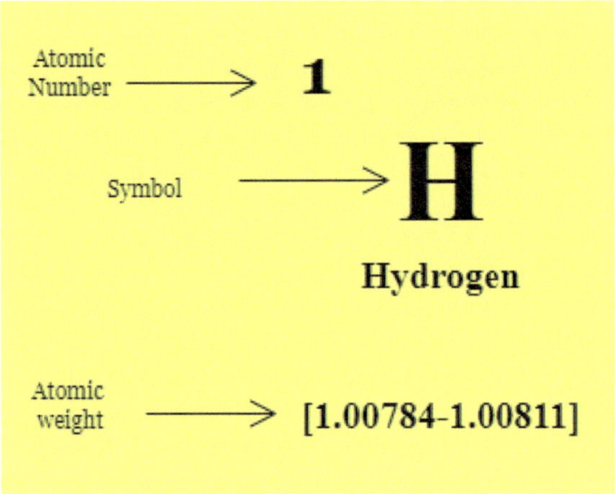

Fig. 4.5 Hydrogen atom symbol with atomic weight in AMU

energy carrier that overcomes many challenges associate with fossil fuel. Hydrogen is represented by letter H and having atomic number 1 in periodic table. Atomic weight of hydrogen is 1.008 AMU. It is the lightest element in the periodic table and has one proton and no neutron in its chemical structure as shown in Fig. 4.5 (Zohuri 2018).

The physio-chemical properties of hydrogen, gasoline, and diesel are shown in Table 4.1. Hydrogen has high energy content per unit mass. One kilogram of hydrogen is equivalent to 3.5 L of gasoline. Hydrogen has wide flammability limit as compare to gasoline and diesel. Hydrogen produces very less exhaust emission forming only water vapour. It has no carbon and sulphur content hence CO, CO_2, SO_x and soot is not produced during combustion which makes hydrogen is a very attractive fuel for IC engine.

Although it has few drawbacks like low ignition energy of hydrogen may cause premature ignition and back flash due to hot spots. In hydrogen engine, back flash is biggest drawback. The storage of hydrogen is a big challenge as it has very low energy density per unit volume. The same fuel tank has lower range with hydrogen as compare to gasoline and diesel because hydrogen has a much lower energy density by volume than conventional fuels such as gasoline or diesel. Gasoline and diesel have energy densities of approximately 33.6 and 35.8 MJ/L, respectively, whereas

Table 4.1 Properties of hydrogen (Zohuri 2018; Faizal et al. 2019; Onorati et al. 2022; Shadidi et al. 2021)

S. No	Properties	Hydrogen	Gasoline	Diesel
1	Chemical formula	H_2	C_4–C_{12}	C_{10}–C_{28}
2	Phase at STP	Gas	Liquid	Liquid
3	Molecular weight (gm/mol)	2.016	114–125	130–285
4	Density (kg/m^3)	0.08375	730–780	820–860
5	Lower heating value (MJ/kg)	120	43.9	42.5
6	Diffusivity (cm^2/s)	0.63	0.08	–
7	Specific heat capacity (Gas) (kJ/kg-k) at 25°	14.304	–	1.5–2
8	Auto-ignition temp (°C)	584	280	180–320
9	Heat of evaporation (kJ/kg) at 1 bar	461	375	–
10	Flammability limit (vol%)	4–75	1.4–7.6	0.6–5.5
11	Minimum ignition energy (MJ)	0.017	0.29	0.24
12	Flame velocity (cm/s)	265–325	37–43	30
13	Adiabatic flame temp. (°C)	2207	2307	2093
14	Vapour pressure (kPa at 20 °C)	234	48–100	–
15	Research octane number	130	92–98	30
16	Stoichiometric AFR	34	14.7	14.5

hydrogen has a much lower energy density of approximately 8.5 MJ/L at ambient temperature and pressure. This means that, for the same volume of fuel, a hydrogen tank contains less energy than a gasoline or diesel tank. As a result, for an equivalent driving distance, the hydrogen tank needs to be significantly larger or pressurized at an extremely high level, and that creates practical difficulties (Hwang and Varma 2014).

Hydrogen can produce from variety of source like conventional, renewable, and nuclear (Kothari et al. 2008). Conventional methods include natural gas reforming, coal gasification, and partial oxidation of hydrocarbons, while renewable methods involve electrolysis powered by solar, wind, and hydro energy. For using hydrogen in IC engine, different types of method are used for injecting the fuel like carburation method, manifold injection, and direct cylinder injection.

4.2 Hydrogen as a Fuel in IC Engine

In current era, hydrogen can be used as an alternative fuel in IC engines for its special characteristics. It is less likely to suffer from engine knock thanks to its wide flammability range and high auto-ignition temperature, which enable high compression, lean burn operations. Moreover, the main by-product of burning hydrogen is water. The pollutants carbon dioxide (CO_2), carbon monoxide (CO), and unburned hydrocarbons (UBHC) are emitted in very small amounts. Hydrogen is an eco-friendly, reliable, and safe energy source found abundant in nature. Hydrogen is frequently used in space programme due to its high energy to weight ratio (Faizal et al. 2019). Hydrogen, with its low spark-energy requirement and wide flammability range, emerges as an excellent candidate for direct use in spark-ignition (SI) engines. Hydrogen has high self-ignition temperature, finite ignition delay, and high laminar flame speed result in a reduced knocking tendency in hydrogen-fuelled internal combustion engine vehicles compared to gasoline-fuelled internal combustion engines. The production of nitrogen oxides (NOx) as a result of high combustion temperatures is still a problem. In compression ignition engine, hydrogen cannot be used directly because auto-ignition temperature of hydrogen (585 °C) is higher than diesel (180–320 °C). So, some modification required engine to avoid knock if we use hydrogen in diesel engine (Gupta 2008). The modification required in injector design as diesel is liquid fuel whereas hydrogen is gaseous fuel (Gomes Antunes et al. 2009). The solenoid operated hydrogen gas injector is used in place of regular diesel injector. The cylinder head required minor modification to accommodate this injector (Saravanan et al. 2007) For use of hydrogen in diesel engine, we can use either diesel as a pilot fuel or spark plug for start of combustion. The glow plug type spark plug is used for start of combustion in diesel engine when only hydrogen as a fuel used (Wozniak et al. 2021; Stępień 2021).

4.3 Hydrogen Production

Hydrogen is a versatile energy carrier with vast potential to transform the way we generate, distribute, and utilize energy. As the world transitions towards a low-carbon economy, hydrogen production is gaining significant attention.

Figure 4.6 shows that at the end of 2022, over 48% of worldwide hydrogen production comes from natural gas, 18% from coal, 30% from oil, and just around 4% from electricity (Kothari et al. 2008). In 2022, electricity had a global average renewable share of approximately 33%, implying that renewable energy accounts for only about 1% of global hydrogen generation (Ishaq et al. 2022).

The classification of hydrogen is mainly based on its production methods and environmental impact related to those processes. Each process uses different sources of energy or feedstocks, thus causing differences in carbon emissions and sustainability. A general overview of the methods for the production of hydrogen along with a basis of classification is presented below. According to the different production method of hydrogen, it can be classified as different names (Ishaq et al. 2022; Rasul et al. 2022). Mostly in literature, the hydrogen production is divided in following categories (Fig. 4.7):

I. Black hydrogen
II. Grey hydrogen

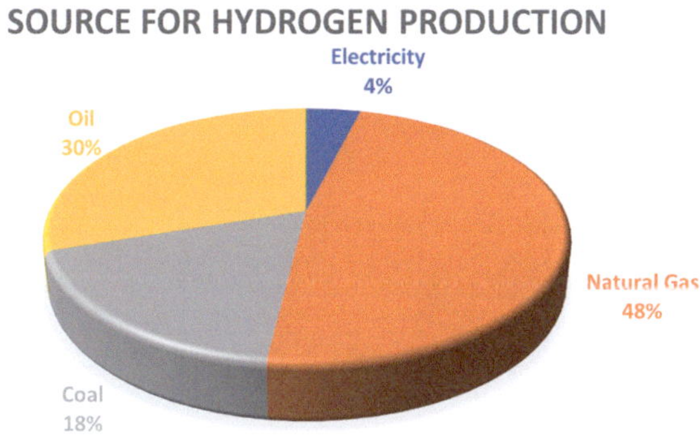

Fig. 4.6 Energy sources for global hydrogen production (Kothari et al. 2008)

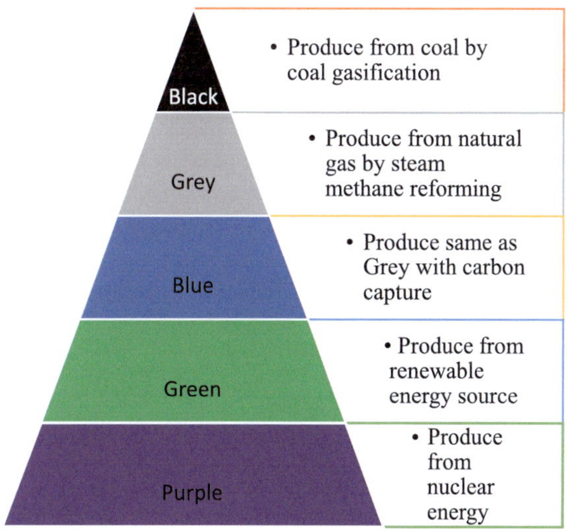

Fig. 4.7 Classification of hydrogen production

III. Blue hydrogen
IV. Green hydrogen
V. Purple hydrogen

4.3.1 Black or Brown Hydrogen Production

Black or brown hydrogen is produced from coal-by-coal gasification. Colours black and brown refer to types of coal bituminous and lignite respectively. By this process, CO_2 produced in large amount. For 1 kg of brown or black hydrogen production, 20 kg of CO_2 is produced and increases the global warming. China has large coal reservoir so it produces hydrogen by coal gasification (Ji and Wang 2021).

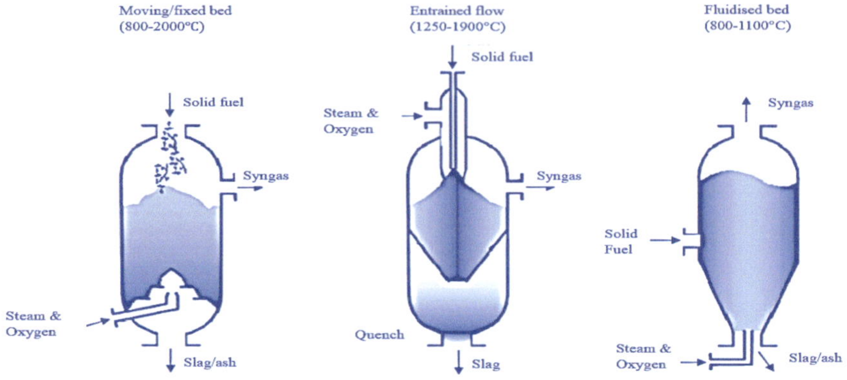

Fig. 4.8 Gasification techniques for brown and black hydrogen production with coal gasification (Kamble et al. 2019)

Coal gasification: Gasification means converting carbon based raw material into syngas by using air, steam, and oxygen. In coal gasification, raw material (e.g. Coal) and waste material converted into valuable product (CO, H_2, CH_4, ash). Purification is required to remove contaminants, particles, and other substances from end product. But due to purification, calorific value is decreased.

These materials are gasified at temperature higher than 900 °C by applying fixed bed gasification, moving bed gasification, and fluidized bed gasification as shown in Fig. 4.8 (Kamble et al. 2019).

4.3.2 Grey Hydrogen Production

Grey hydrogen is primarily created through steam methane reforming (SMR) or coal gasification. SMR is the process of reacting methane (natural gas) with steam under high pressure to create hydrogen, carbon monoxide, and a little quantity of CO_2. Coal gasification is the process of reacting coal with oxygen and steam to create hydrogen and CO_2.

Unlike blue hydrogen, grey hydrogen generation does not use carbon capture, utilization, and storage (CCUS) technology. This implies that all of the CO_2 created during the process is discharged into the atmosphere, getting worse the environmental impact.

Grey hydrogen is mainly produced via steam methane reforming (SMR) or coal gasification, both of which are well-established and mature technologies. For many years now, these techniques have been widely employed, and this means their technical challenges are well-defined and can be managed. The production process is relatively inexpensive because it utilizes abundant and low-cost fossil fuels, thus meeting a significant portion of the world's industrial hydrogen requirements with particular emphasis on petrochemicals and ammonia making. However, the grey hydrogen also has other notable drawbacks that include its environmental impact. The process emits quite a substantial amount of CO_2 approximately 830 million metric tonnes annually hence significantly contributing to greenhouse gas emissions. Moreover, it heavily relies on finite fossil fuels hence less sustainable compared to renewable hydrogen production methods. High CO_2 emissions also cause air pollution thereby negatively affecting both the environment and human health. Given that global efforts increase in mitigating carbon footprints as well as fighting against climate change these weaknesses become more important (Saha et al. 2024).

4.3.2.1 Coal Gasification

Gasification of coal involves several intricate steps. The core of the process is the reaction of coal with a controlled amount of oxygen and steam at high temperatures, which prevents complete combustion. This partial oxidation results in the production of syngas, a mixture of hydrogen, carbon monoxide, carbon dioxide, and other trace gases (www.HydrogenAssociation.org). In this method, following processes are carried out shown in Fig. 4.9 (Jayaprabakar 2024).

The chemical reaction in this process is as follows:

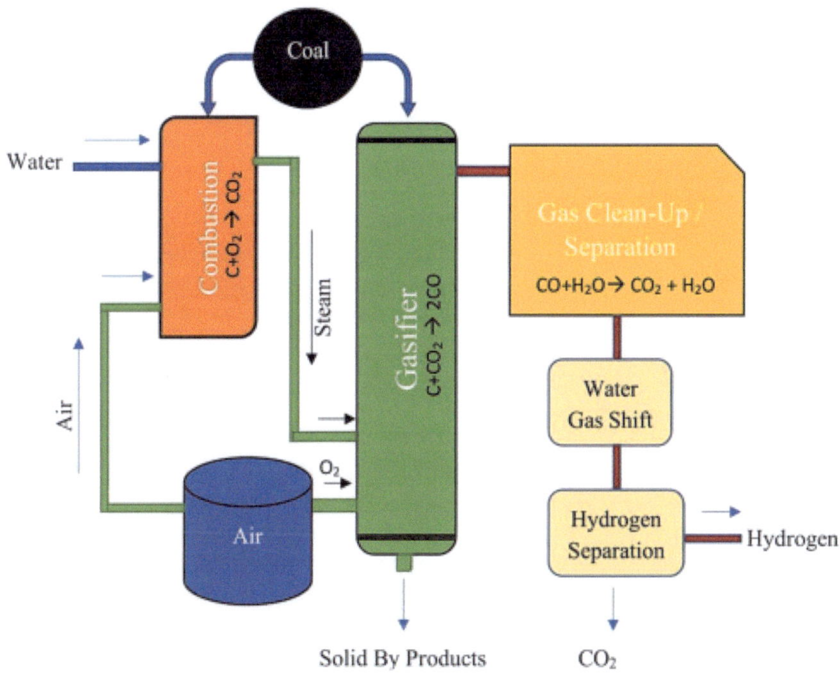

Fig. 4.9 Hydrogen production by coal gasification (Jayaprabakar 2024)

$$\text{Partial oxidation: C} + \left(\frac{1}{2}O_2\right) \rightarrow CO \tag{4.3.1}$$

$$\text{Water gas shift reaction: } CO + H_2O \rightarrow CO_2 + H_2 \tag{4.3.2}$$

The main disadvantage of this method is that it produces carbon dioxide as well as harmful pollutants like sulphur dioxide, nitrogen oxide, and particulate matter (www.HydrogenAssociation.org).

4.3.3 Blue Hydrogen Production

Blue hydrogen is mostly produced from natural gas through steam reforming, in which natural gas and heated water are combined to form steam. The output is hydrogen; however, it also produces carbon dioxide as a by-product. Blue hydrogen is commonly referred to as "low-carbon hydrogen", but the steam reforming process does not eliminate the production of greenhouse gases. Blue hydrogen is slightly less hazardous to the environment than grey hydrogen. Blue is created in the same way as grey, but the carbon dioxide is captured and stored rather than emitted into the atmosphere.

The methods of blue hydrogen productions are:

I. Natural gas reforming.
II. Partial oxidation of hydrocarbon.

4.3.3.1 Natural Gas Reforming

Gas reforming having three methods to produce hydrogen and syngas.

I. Steam reforming of methane (SRM).
II. Dry reforming of methane (DRM).
III. Bi-reforming of methane (BRM).

I. Steam Reforming of Methane

SRM is currently one of the most important and commonly used technologies for producing hydrogen. SRM is produced around 50% of the world's total hydrogen. This endothermic catalytic process operates at temperatures above 500 °C and pressures of up to 35 bars. Nickel acts as a catalyst in this process. The high H/C ratio of fossil fuels makes them suitable candidates for reforming, with natural gas serving as the most common input. However, heavier hydrocarbons, like naphtha can be handled satisfactorily (Kothari et al. 2008). A considerable amount of the steam created during the steam reforming process exceeds the system's requirements. This excess steam can be exported directly or transformed into electricity using specially constructed extra facilities. The basic reaction occurs in two steps:

$$CH_4 + H_2O \rightarrow CO + 3H_2 \quad (4.3.3)$$

$$CO + H_2O \rightarrow CO_2 + H_2 \quad (4.3.4)$$

Equation 4.3.1 is primary steam reforming process and Eq. 4.3.2 is water gas shift reaction. The syngas obtained typically has H_2 ratio of 3:1. A significant disadvantage of the steam reforming process is the substantial emission of carbon dioxide (CO_2) into the atmosphere. This is primarily due to the reliance on fossil fuels both as feedstock in the manufacturing process and as the source of heat required for the reaction. Consequently, the environmental impact of steam reforming is considerable, contributing to greenhouse gas emissions and climate change. Figure 4.10 shows a block diagram of hydrogen production from natural gas reforming.

II. Dry reforming of methane

DRM is a method for producing syngas through the reaction of methane and carbon dioxide at a temperature above 750 °C. DRM effectively utilized CH_4 and CO_2, helping to mitigate their greenhouse effect. The primarily reaction of DRM is:

$$CH_4 + CO_2 \rightarrow 2CO + 2H_2 \quad (4.3.5)$$

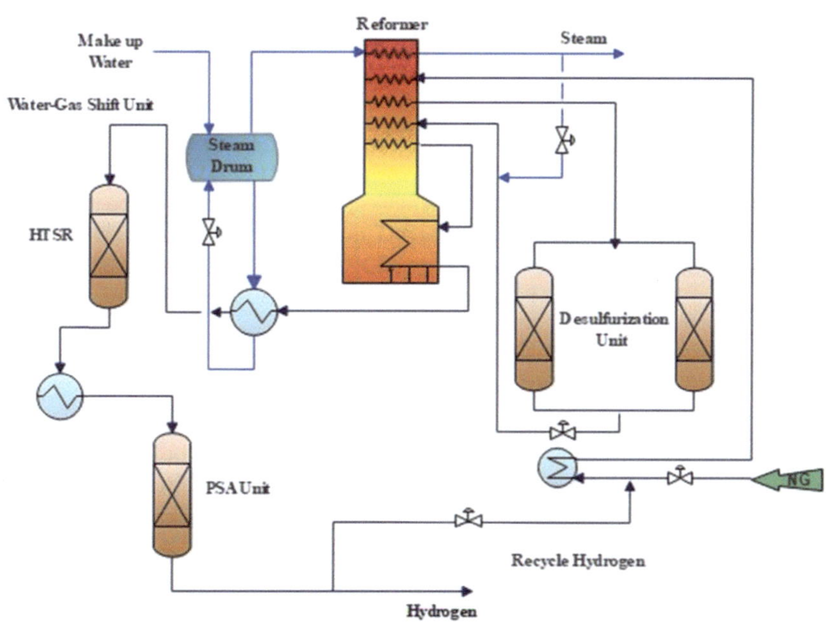

*NG: Natural Gas, HT-SR: High Temperature Steam Reforming

Fig. 4.10 Hydrogen production from SRM method (Taji et al. 2018)

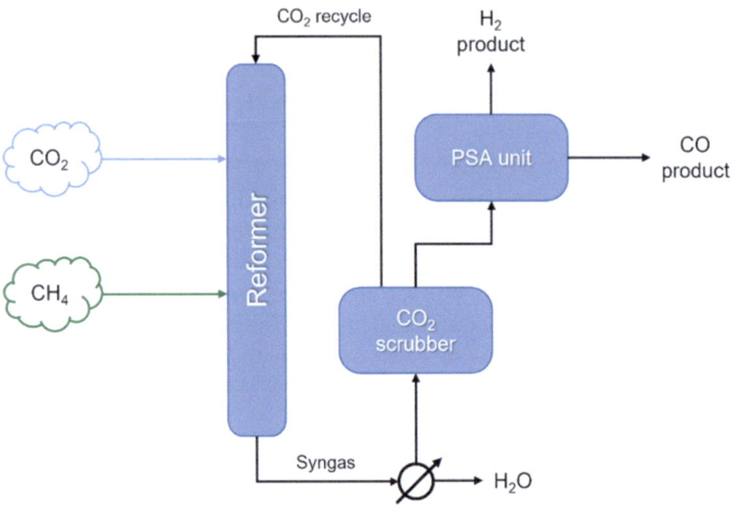

Fig. 4.11 Syngas production by dry reforming method (Medeiros et al. 2022)

The syngas obtained typically has H_2 ratio of 1:1. The major drawback of DRM is high operating temperature. Due to this temperature, coke formation and catalysts sintering problem is occurs. To reduce this problem, noble-based metal catalysts is used but cost of noble-based catalysts is high as compared to Ni-based catalysts (Taji et al. 2018). Figure 4.11 shows dry reforming of methane for producing syngas (Medeiros et al. 2022).

III. Bi-reforming of methane

It is a catalytic process that transforms methane (CH_4) into syngas, which is a blend of carbon monoxide (CO) and hydrogen (H_2). This conversion is achieved by utilizing a combination of carbon dioxide (CO_2) and steam (H_2O). The general chemical reaction for bi-reforming is as follows:

$$3CH_4 + CO_2 + 2H_2O \rightarrow 4CO + 8H_2 \quad (4.3.6)$$

For efficient hydrogen and syngas production, BRM requires high temperature (above 800 °C). The flexibility of methane bi-reforming is due to its ability to utilize natural gas with high levels of CO_2 such that there would be no need for CO_2 separation and making use of CO_2 from flue gases effective. This technology produces synthesis gas which has an H_2/CO ratio of approximately 2:1, a composition preferred for industrial processes including Fischer–Tropsch synthesis or methanol production. The addition of steam in the reaction helps control carbon by reducing deposition on catalysts, which is typical during dry reforming. Bi-reforming is environmentally friendly since it uses up CO_2 thus assisting in mitigating carbon emissions.

One drawback of bi-reforming methane is that it requires much energy. The reaction taking place in it is extremely endothermic and therefore requires high energy

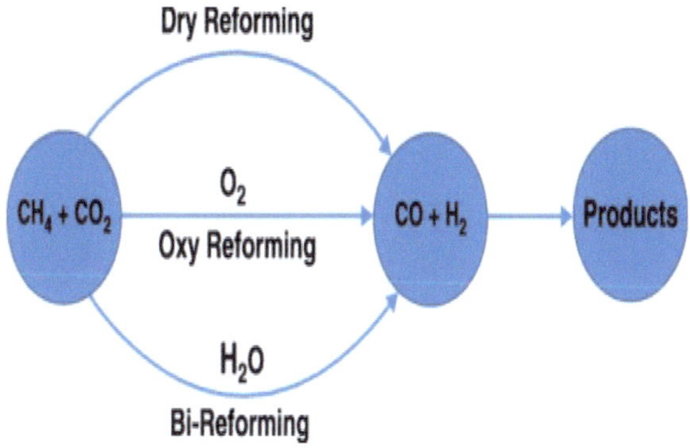

Fig. 4.12 Syngas production by BRM (Kumar et al. 2015)

inputs to maintain the reaction temperature. It has reduced carbon formation, but still there is the challenge of deactivation, which arises from sintering or fouling among other forms, thus needing frequent repairs and replacements. Moreover, there is a difficulty in balancing CO_2 and H_2O inputs for obtaining desired syngas composition with respect to process control. Moreover, the suitability of existing infrastructure worsens the situation. The extreme temperatures required for my reforming processes, along with the need for specific and costly catalysts, limit their widespread adoption in commercial markets (Kumar et al. 2015). Figure 4.12 shows syngas production by BRM method.

4.3.3.2 Partial Oxidation of Hydrocarbon

The partial oxidation of hydrocarbons, such as methane, involves a process where the hydrocarbon reacts with some oxygen but not enough to enable complete combustion. The reaction produces a stream containing equal amounts of carbon monoxide and hydrogen commonly referred to as synthesis gas or syngas. The reaction is exemplified below:

$$CH_4 + 0.5O_2 \rightarrow CO + 2H_2 \tag{4.3.7}$$

In this process, methane undergoes partial oxidation using oxygen to give carbon monoxide (CO) and hydrogen (H_2). This means that when it occurs, heat is released.

Various industrial processes rely on partial oxidation which is used primarily for manufacturing hydrogen in fuel cells and generating syngas needed in chemical synthesis. As opposed to complete oxidation which gives rise to carbon dioxide and water, incomplete combustion provides an easier method of producing hydrogen and carbon monoxide. Production of high temperatures essential for the process can be done through techniques like plasma-assisted catalysis (PAC) which increase efficiency and output of the process. Figure 4.13 shows partial oxidation method of hydrogen production (Chao et al. 2008).

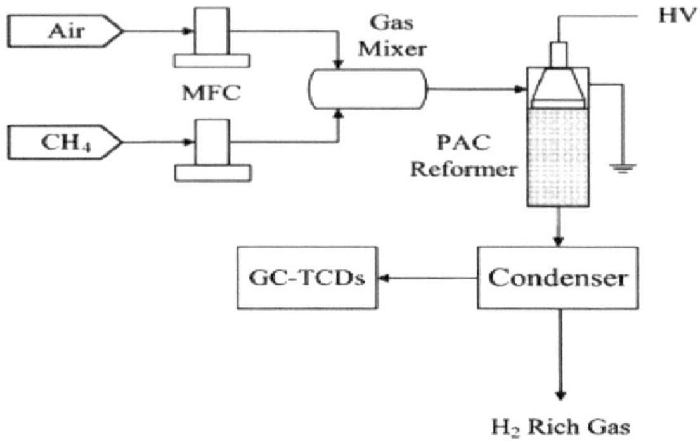

Fig. 4.13 Partial oxidation method of hydrogen production (Chao et al. 2008)

There are several reasons why the partial oxidation of methane is an ideal strategy to hydrogen production. For example, it can achieve high efficiency especially when coupled with plasma-assisted catalysis thereby leading to increased rates of conversion for methane to carbon monoxide and hydrogen. This means that a lot of hydrogen is produced in an efficient way while using a small amount of methane. Moreover, the process is exothermic such that it produces internal heat thus reducing the need for external energy sources which improves overall energy utilization. The system is designed compactly hence enabling fast start-up and response times making it suitable for on-demand hydrogen production as well as portable applications. Using methane as a feedstock capitalizes on its abundance and low cost, making it possible to exploit effectively existing natural gas infrastructure and resources. In addition, this syngas made up of carbon monoxide and hydrogen can be used directly as clean fuel or serve as intermediates towards other chemicals manufacturing processes.

Though methane partial oxidation has a number of advantages, it also comes with its own fair share of disadvantages. The major problem is that carbon monoxide is produced which is deadly and can therefore hinder pure hydrogen utilization in such applications. In order to tackle this concern, more steps may have to be taken in separation or conversion of carbon monoxide. Another downside that must be dealt with is catalyst deactivation whereby the effectiveness of catalysts diminishes over time due to coking or other reasons; hence, replacement and regeneration are required periodically. Moreover, if not adequately supplied with oxygen during the process, methane pyrolysis will occur leading to formation of soot, consequently contaminating the catalyst further and inhibiting the reaction. When operating such reactors at high temperatures required by the process, it can prove difficult due to need for careful control in order not only to avert destruction but also ensure efficiency. On top of thermodynamic barriers influencing efficacy especially at lower temperatures, optimizing reaction conditions becomes pivotal as high conversion rates and yields are desired.

4.3.4 Green Hydrogen Production

Green hydrogen is a clean, sustainable, and versatile energy carrier produced using renewable sources of energy. It is generated through the electrolysis of water, where electricity from renewable sources such as solar, wind, or hydropower splits water into hydrogen and oxygen. This process produces hydrogen without emitting greenhouse gases, which makes it different from "grey" and "blue" hydrogen, which involve fossil fuels and carbon capture. For the decarbonization of industries including energy storage, heavy manufacturing, and transportation, green hydrogen is needed. Green hydrogen helps achieve net-zero emissions and improves energy security by combining electrolysis with renewable electricity (Dash et al. 2023). Green hydrogen production is shown in Fig. 4.14.

Green hydrogen can be produced from mainly three different methods:

I. Electrolysis by solar energy.
II. Electrolysis by wind energy.
III. Electrolysis by hydro energy.

Electrolysis: Electrolysis is a suitable method of generating hydrogen which emits or produce zero carbon during production. Although the input energy for electrolysis process should be either from renewable sources. In this process, the water is break down into hydrogen and oxygen with the help of electricity. The hydrogen is collected on cathode and oxygen on anode. This chemical reaction of electrolysis is occurring in electrolyser. The selection of the electrolyser also depends on the planned production capacity. Electrolysers can be chosen in sizes ranging from small to large, depending on the need (Acar and Dincer 2014). The electrolysis process is shown in Fig. 4.15.

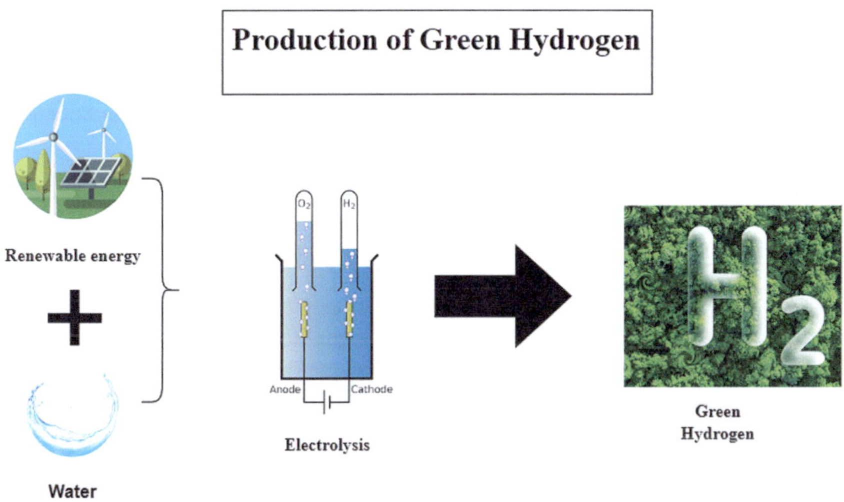

Fig. 4.14 Green hydrogen production

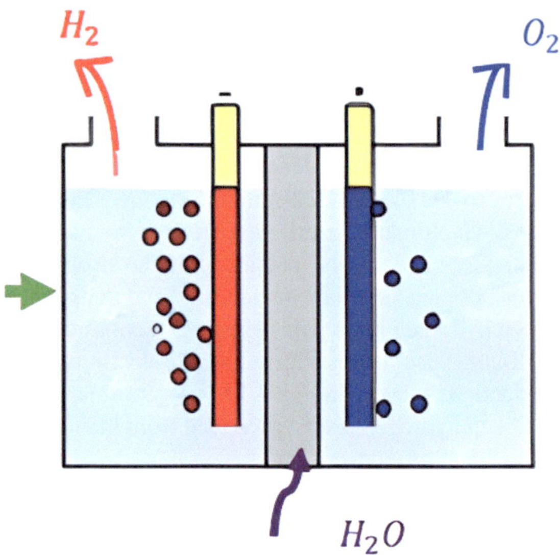

Fig. 4.15 Electrolysis process of hydrogen production (Ajeeb et al. 2024)

The electrolysis process is further classified on the basis of key process used including alkaline water electrolysis (AEL), proton exchange membrane (PEM), anion exchange membrane (AEM), and solid oxide electrolysis (SOE). The diagram of different technologies is shown in Fig. 4.15.

PEM: Proton exchange membrane (PEM) electrolysis is an efficient process involving a solid polymer electrolyte to split water into hydrogen and oxygen. It works at acidic conditions, has fast response capabilities to variable renewable energy feeds, and delivers high purity hydrogen but requires expensive noble metal catalysts.

AEM: AEM electrolysis combines features of PEM and AEL. This is an electrolysis method using a solid membrane in an alkaline environment to produce hydrogen. It avoids expensive catalysts and has potential for low-cost production, though it needs further development to match the performance of other methods.

AEL: AEL uses a liquid alkaline solution, such as KOH, as the electrolyte; in it, hydrogen is produced due to hydroxide ion migration. The process is economical and quite durable but less efficient than PEM with slower response time.

SOE: Solid oxide electrolysis (SOC) is a high-temperature technology using a solid ceramic electrolyte to convert steam into hydrogen and oxygen. Its high thermal efficiency makes it very suitable for industrial applications, but the extreme operating conditions pose challenges in material durability and system stability.

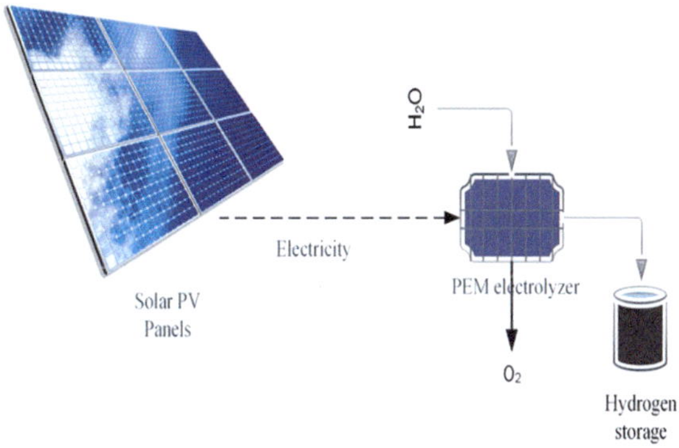

Fig. 4.16 Solar PV-based hydrogen production (Ishaq and Dincer 2021)

4.3.4.1 Electrolysis by Solar Photovoltaic

Generation of hydrogen by solar energy is an emerging trend for future production of fuels. The abundant sunlight is available in Middle East and Asian countries. The sunlight is used to generate clean hydrogen through numerous means like photovoltaic cells with the electrolysis of water. Another technology is photoelectrochemical (PEC) water splitting, which directly utilizes solar energy to dissociate water molecules using specialized photoelectrodes; during this, solar-generated electricity splits water molecules into hydrogen and oxygen (Ture 2007).

Photovoltaic solar panel absorbs radiation from sun and generates electrical energy. Generated electricity from solar panel further used in electrolysis process and then hydrogen will produce. Ishaq et al. done a software analysis, in which 20 solar PV panels are considered, with a solar irradiance of 800 W/m^2 and an assumed efficiency of 15%. The electricity produced by the solar PV panels is DC, which is suitable for direct use by the PEM electrolyser. The PEM electrolyser operates at 80 °C and 1 bar of pressure, utilizing the DC electrical output from the PV panels to split water into hydrogen and oxygen. The overall energy and exergy efficiencies were 16.95 and 17.45%, respectively. The hydrogen production rate was 2.26 g/s in their analysis. Figure 4.16 shows the solar PV-based hydrogen production (Ishaq and Dincer 2021).

4.3.4.2 Electrolysis by Wind Energy

The wind energy has a great potential to generate electricity without zero carbon emission. The hydrogen can be produced by electrolysis through electricity generated by wind energy especially for distributed system. The cost of wind turbine is very high; hence, it costs almost 6–10 times of fossil fuel alternative. So, for making this technique viable, need of optimization requires for electricity generation (Acar and Dincer 2014). A study has conducted by Mustafa et al. for different regions of Turkey and different hub heights ranging from 50–100 m. They found that hydrogen

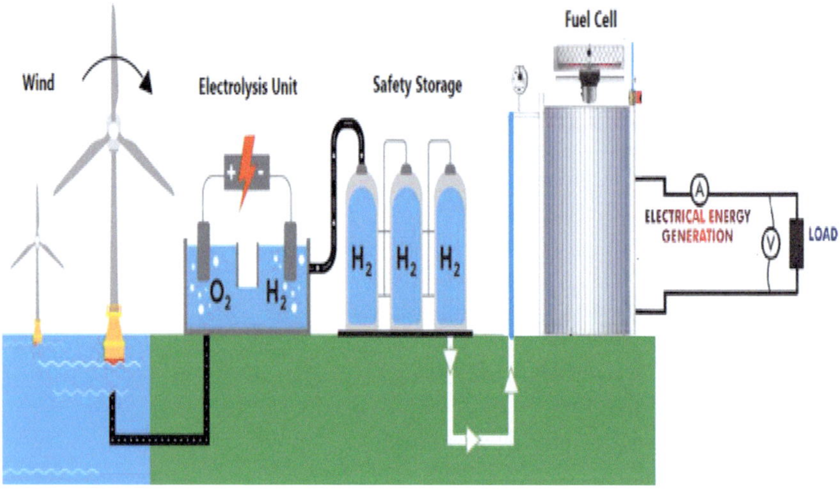

Fig. 4.17 Hydrogen production by wind energy (Benghanem et al. 2023)

production using a 1300 kW wind energy conversion system is depend upon hub height and location varied from 1665.24 kg H_2/year in Nigde at 50 m hub height to 6288.59 kg H_2/year in Pinarbasi at 100 m hub height (Genç et al. 2012). The schematic diagram of hydrogen production method by wind energy is shown in Fig. 4.17 (Genç et al. 2012).

4.3.4.3 Electrolysis by Hydro Energy

Hydrogen production from hydropower energy is an increasingly promising area to contribute significantly to a clean energy future. The method usually breaks water molecules with electricity by electrolysis and produces hydrogen gas. The process stands as a dependable and less polluting option compared with the traditional method of extracting hydrogen from fossil fuel-based sources. Figure 4.18 showed different route of hydro energy generation and process of hydrogen generation (Ishaq et al. 2022).

4.3.5 *Purple Hydrogen Production*

Purple hydrogen, or hydrogen produced using nuclear energy, takes advantage of the high temperatures found in nuclear reactors to produce hydrogen through thermo-chemical processes. The fission of uranium atoms produces heat, which turns into steam, which powers turbines to create electricity, and which is the source of nuclear

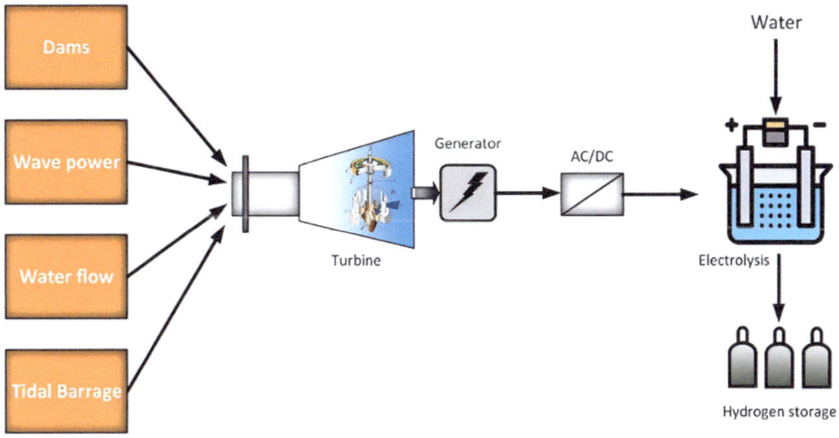

Fig. 4.18 Electrolysis by hydro energy (Ishaq et al. 2022)

energy. Nuclear power facilities do not directly burn fuel, in contrast to fossil fuels, and as a result do not produce greenhouse gas emissions.

The fission reaction is regulated inside the reactor of nuclear power plants, offering a dependable source of heat. Fusion reactions are not used despite being potentially more efficient since they are expensive and difficult to maintain under the necessary circumstances.

Hydrogen can be produced using the thermal energy produced by the fission reaction in addition to electricity (Ishaq et al. 2022). Typical techniques consist of:

- Turbine electricity to fuel cells: Fuel cells may produce hydrogen by using the generated electricity.
- Thermochemical water splitting: The process of splitting water molecules using heat energy.
- Steam reforming: Steam reforming is the process of turning hydrocarbons into hydrogen by applying heat.
- Water electrolysis: It is the process of splitting water into hydrogen and oxygen using electricity produced by nuclear power.

Nuclear energy is a flexible source for producing hydrogen and electricity thanks to these techniques.

4.4 Hydrogen Injection Techniques in IC Engine

In the search for clean, sustainable energy sources, hydrogen has drawn a lot of attention due to its clean burning nature and renewable nature. Its use in internal combustion engines (ICEs) offers a viable substitute for traditional fossil fuels, resolving the fuel problem as well as worries about pollution to the environment.

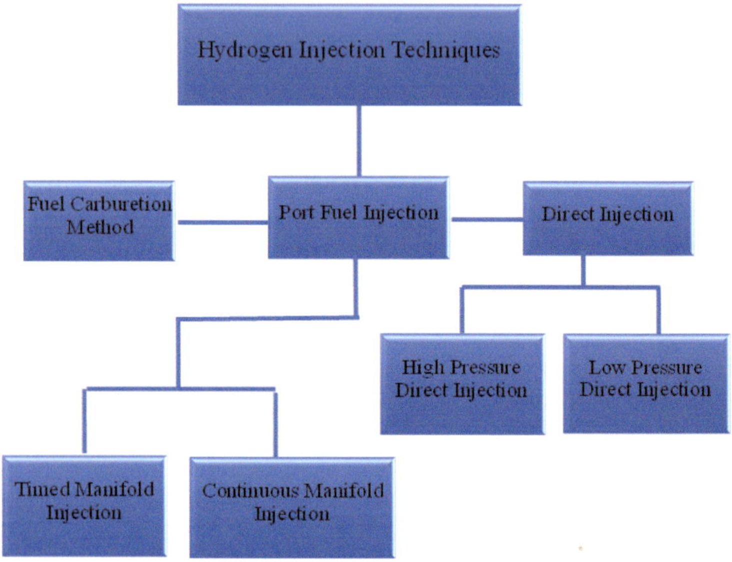

Fig. 4.19 Hydrogen injection techniques

There are several injection strategies available for integrating hydrogen into internal combustion engines (ICE) which are shown in Fig. 4.19.

4.4.1 *Fuel Carburation Method*

The hydrogen fuel carburation injection technique involves introducing hydrogen gas into the air intake of an internal combustion engine with the help of gas carburettor. There are several benefits of this method. First was it being oldest and frequent used technique especially in SI engine. The gas carburettor, which used in gasoline engines, can be used to facilitate the hydrogen fuel gasification process. Secondly, as compared to other techniques, it does not require a high hydrogen supply pressure.

A schematic diagram of a hydrogen fuel carburation method of injection technique is shown in Fig. 4.20. In this system, air and hydrogen is mixed in a gas carburettor and supplied this mixture to inlet manifold of an IC engine. The quantity of hydrogen is controlled by a regulator according to the load condition and the throttle position. For proper operation, some engines required water addition (Dash et al. 2022).

This method of injection has several drawbacks like it produces less power. The problem of backfire is serious concern from the safety point of view. It has chances of pre-ignition and knocking. If pre-ignition occurs, the flame can reached in the inlet manifold from combustion chamber through inlet valve and resulted in backfire. In case of carburetted hydrogen injection engine, it is quite serious. Because the inlet manifolds have fuel–air mixtures in flammability range. Hence, special care must be

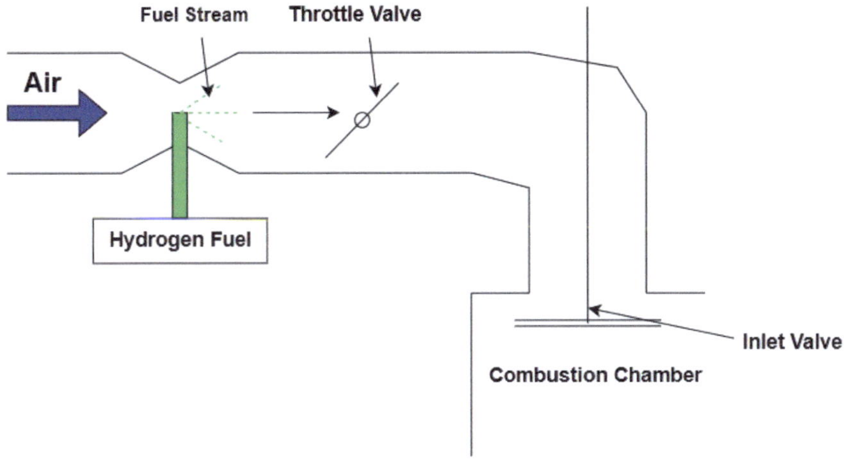

Fig. 4.20 Fuel carburation method

taken to avoid backfire or pre-ignition in this technique. Due to a number of serious disadvantages, carburetion is not a viable technique for hydrogen injection in internal combustion engines.

4.4.2 Inlet Manifold/Port Fuel Injection

The port injection method is a technique used to introduce the hydrogen fuel directly into the intake manifold of an internal combustion engine. The schematic diagram of port fuel injection is shown in Fig. 4.21.

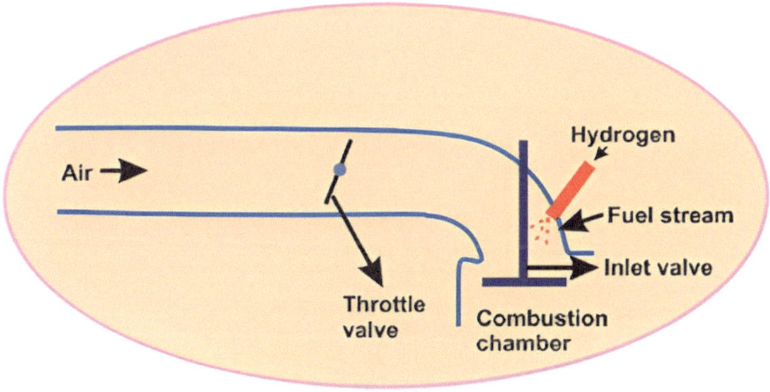

Fig. 4.21 Port fuel injection (Hosseini et al. 2023)

In this technique, the inducted volume of air per cycle is kept same. To control engine output, the quantity of fuel has varied. The fuel quantity is controlled either by change the injection pressure or the injection duration. The injection duration of is controlled by controlling the signal pulse to the injector.

Hydrogen is injecting by means of electrical or mechanical injector. This injection technique is reduced the chances of backfire. The reason behind this phenomenon is that in the starting of intake stroke, only air is supplied, which leans the residual gases and quenches the hot spots. Figure 4.22 shows the manifold injection mechanism.

There are two methods for inlet manifold injection: first one is continuous manifold injection and second is timed manifold injection. In continuous manifold injection, the hydrogen is supplied continuously with air in intake manifold at a steady rate, providing a constant supply throughout the engine's operation. In the timed manifold injection, the hydrogen is injected at specific intervals into the intake manifold during the engine's operation.

The comparison of timed manifold injection verses continuous manifold injection is shown in Table 4.2.

Port fuel injection (PFI) controls pre-ignition by delivering fuel into the intake port before the intake valve opens, so that a homogeneous air–fuel mixture can form there, which lowers hotspots that could lead to pre-ignition. Evaporation of the fuel cools the intake charge, lowering combustion temperatures and minimizing the risk of spontaneous ignition. Moreover, since the fuel does not come in direct contact

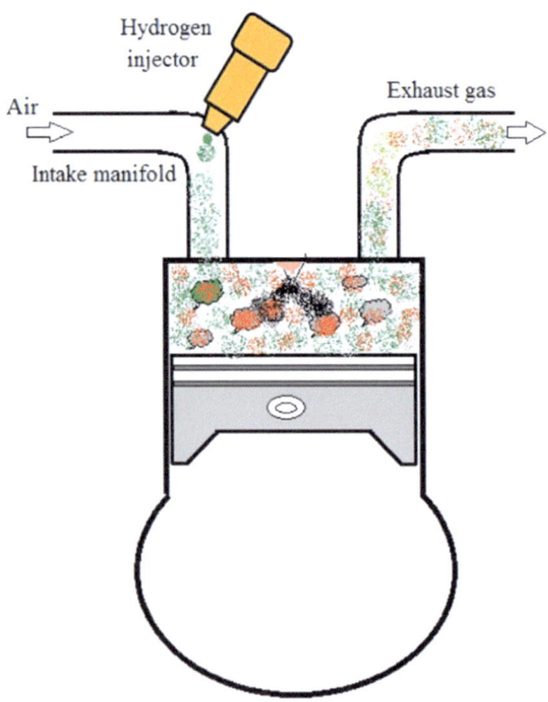

Fig. 4.22 Manifold injection in diesel engine

Table 4.2 Comparison of timed manifold verses continuous manifold injection

Features	Timed manifold injection	Continuous manifold injection
Injection timing	At specific interval	Continuous
Cost	High	Low
Complexity	More complex, require precise time control	Less complex, but not precise
Efficiency	Higher due to precise control	Lower due to inconsistency
Emission	Can lower with optimized condition	Higher due to less control
Power output	Higher	Lower

with the surfaces of the hot combustion chamber, the risk of local heating causing premature ignition is reduced. This makes PFI effective in pre-ignition.

4.4.3 Direct Cylinder Injection

This technique injects the hydrogen fuel directly into the combustion chamber after the inlet valve completely closed during the compression stroke. As during compression stroke pressure inside the cylinder is high, so in this technique, hydrogen is injected at higher pressure as compare to PFI and fuel carburation method. This approach encourages rapid diffusion and near-instantaneous mixing of hydrogen with air. Figure 4.23 shows a schematic diagram of direct cylinder hydrogen injection system. It can be seen from the diagram that during fuel injection, inlet valve has closed.

Direct cylinder injection has classified into two techniques:

1. Low-pressure direct injection (LPDI),
2. High-pressure direct injection (HPDI).

A low-pressure direct injector injects fuel when the intake valve closes and there is low pressure inside the cylinder means just after the starting of compression stroke. The high-pressure direct injector delivers fuel at higher pressure at the end of the

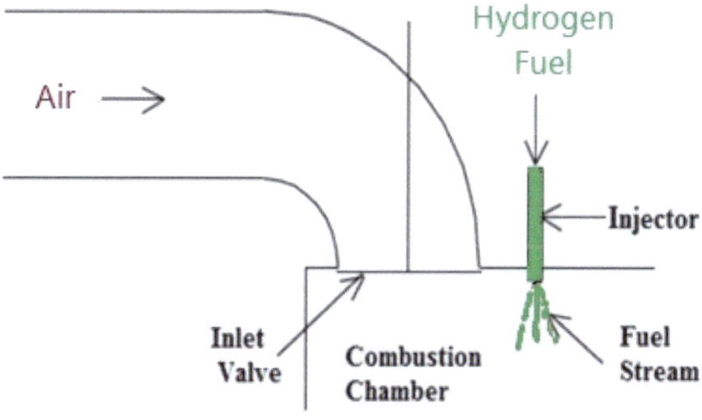

Fig. 4.23 Direct cylinder injection

compression stroke (www.erpublication.org). The comparison of LPDI and HPDI is shown in Table 4.3.

The low-pressure direct injection (LPDI) systems pump fuel directly into the combustion chamber at relatively low pressures, typically between 10 and 200 bar. This method is often used for injecting gaseous fuels like natural gas or hydrogen. LPDI have some advantages as follows:

- Simplified Injection System: Because of the decreased pressure requirements, the injection system's design is simpler and less expensive.
- Improved Mixing: When designed correctly, LPDI can provide good fuel–air mixing, resulting in higher combustion efficiency.
- Lower NOx Emissions: Because LPDI systems provide greater control over the air–fuel mixture and combustion temperature, they can assist reduce NOx emissions.
- Energy Efficiency: LPDI systems use less energy to compress the fuel than high-pressure systems.

The high-pressure direct injection (HPDI) systems pump fuel directly into the combustion chamber at high pressures, usually between 500 and 2500 bar or even higher. This method applies to both liquid fuels (such as diesel) and gaseous fuels (such as hydrogen and natural gas). HPDI has some advantages as follows:

- Superior atomization: The high injection pressure guarantees that the fuel is atomized perfectly, producing fine droplets that mix more effectively with air, resulting in more complete combustion.
- HPDI systems can offer larger power outputs because of improved air–fuel mixing and combustion efficiency

Table 4.3 Comparison of LPDI versus HPDI

Feature	Low-pressure direct injection	High-pressure direct injection
Injector pressure	Low	High
Injector design	Simpler	Complex
Fuel atomization	Coarse	Fine
Combustion efficiency	Less due to coarse fuel	High due to homogeneous mixture formation
Emission	Generally higher	Lower due to better combustion
Power output	Lower	Higher due to complete combustion
Robustness	More robust	Less robust, require high reliability materials
Cost	Lower	Higher
Safety	Higher due to lower pressure	Lower due to high pressure
Applications	Preferred where cost and simplicity are on priority, old vehicles	Preferred where high performance required. New modern vehicles

Table 4.4 Comparison of hydrogen injection techniques

Parameter	Carburetion method	Port fuel injection	Direct injection
Injection location	H_2 mix with air in carburettor	H_2 is injected into the intake manifold before intake valve	H_2 is injected direct into combustion chamber
Risk of pre-ignition	High	Moderate	Low
Backfire risk	High	Moderate	Minimal
Control over air–fuel ratio	Poor	Moderate	Excellent
Efficiency	Lower	Moderate	High
Complexity	Simple	Moderate	High

- Lower particulate emissions: Finer fuel droplets and more complete combustion produce lower particulate emissions.
- HPDI systems can improve engine efficiency by improving the combustion process.

The comparison of hydrogen injection techniques is shown in Table 4.4.

4.5 Real-World Application of Hydrogen

Hydrogen has been increasingly used in the vehicle sector as a cleaner and more sustainable alternative fuel source compared to the more traditional fossil fuels. This is primarily through the hydrogen fuel cell electric vehicles (FCEVs) application, where hydrogen gas powers electricity through a fuel cell. This electricity drives an electric motor, which, without any emission except for water vapour, makes FCEVs an eco-friendly approach to reducing greenhouse gas emissions. Automakers such as Toyota (Mirai), Hyundai (Nexo), and Honda (Clarity Fuel Cell) have developed hydrogen-powered passenger cars that can travel long distances of about 500–600 km with refuelling times of just 3–5 min, making them a practical solution for daily transportation needs (https://www.toyota.com/mirai/).

Public transport systems are also using hydrogen-fuelled buses to lower the extent of urban air pollution and reduce carbon emissions. In freight and logistics, companies, such as Nikola Motors and Hyundai, are investing in fuel cell trucks using hydrogen fuel cells, thus permitting long distances and fast refuelling according to the demanding requirements of heavy-duty applications (https://hydrogencouncil.com/en/path-to-hydrogen-competitiveness-a-cost-perspective/).

Maritime transport is also gaining hydrogen technology. Ferries and cargo ships such as the Hydro Ville in Belgium are using hydrogen fuel cells to reduce emissions in coastal and inland waterways (https://www2.deloitte.com/cn/en/pages/about-deloitte/articles/pr-fcevs-to-be-cheaper-to-run-than-bevs-and-ice-vehicles-within-10-years.html).

4.6 Challenges with Hydrogen in IC Engine

Currently, hydrogen is a promising alternative fuel for IC engine due to its ability to reduce greenhouse gas emission and other pollutants produce by IC engine. However, the use of hydrogen in ICEs raises numerous serious technological and operational challenges like backfire and knock, storage and handling of hydrogen, material compatibility and durability, lubricant issue, lower efficiency, high pressure required, injector design challenges, NO_X emission, and pre-ignition (Sun et al. 2023). These all challenges are discussed in detailed as follows:

Backfire

Backfire in hydrogen engines is the unwanted incident in which an unburned hydrogen-air combination ignites prematurely due to high temperatures and pressures within the cylinder, allowing the flame to escape into the intake system. This incident can reduce volumetric efficiency and cause damage to the intake and fuel injection systems. The higher burn rate of hydrogen in comparison to gasoline and diesel, as well as the lower minimum ignition energy, are the fundamental causes of backfire. Backfire can be caused by faulty intake valve and fuel injection timing, a high equivalency ratio, excessive residual exhaust gas, a sluggish combustion rate within the cylinder, and improper hydrogen distribution around the intake valve. These issues can be addressed by modifying the control technique, but totally eliminating backfire requires switching from PFI to DI engines, which are more resistant to this occurrence (Gao et al. 2022). Figure 4.24a shows the backfire phenomenon in IC engine, and in Fig. 4.24b with the help of flame arrestor, this phenomenon can be prevented.

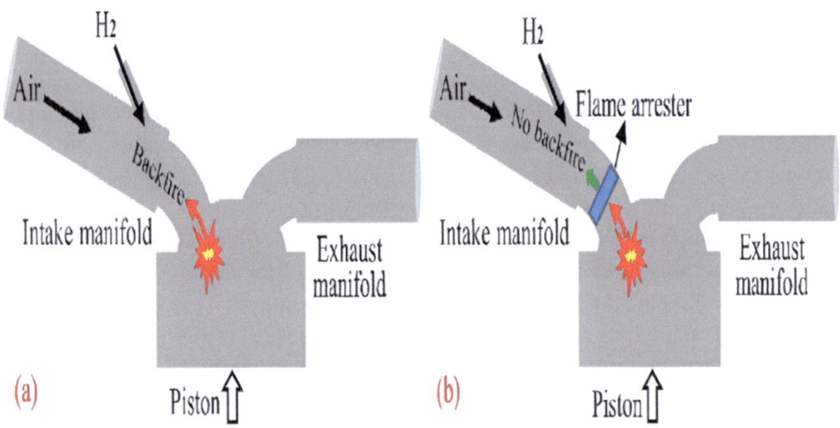

Fig. 4.24 **a** Backfire phenomenon, **b** flame arrester (Sun et al. 2023)

Storage and Handling

Storage and transportation of hydrogen is another problem because of its low energy density (Edwards et al. 2007). The hydrogen requires a big storage space to offer sufficient driving range for automobiles. For storing sufficient amount of hydrogen, a high pressure (around 700 bar) or cryogenic temperature (below −253 °C) is required. Furthermore, its low density requires the use of high-pressure storage systems, which are both costly and hard to implement (Hosseini and Butler 2020).

Material Compatibility and Durability

Hydrogen's strong diffusivity can cause material embrittlement in engine components. It can cause cracks and failures of material. The usage of specialized materials and coating is necessary that can endure hydrogen exposure without cracking. The hydrogen engines must be constructed to manage the unusual combustion properties of hydrogen, such as its increased flame velocity and diffusivity, which need precise control over the fuel–air mixture and ignition timing (Dash et al. 2022).

Lubrication Issues

The absence of lubricating qualities of hydrogen might cause damage to the injector nozzle's sealing surface. Specialized injector designs are necessary to assure durability and reliability.

Lower Efficiency

Hydrogen-powered ICEs typically attain approximately 20–25% efficiency, but hydrogen fuel cells (HFCs) can achieve efficiencies of 60% or more. To give a competitive driving range, cars must have a substantially larger fuel storage tank due to their reduced efficiency (Hosseini and Butler 2020).

Injector Design Challenges

Hydrogen ICE injectors needs to handle high pressures, requires a specialized material to resist embrittlement. The size and design of injector must be very small to fit into the engine head, which normally has four valves per cylinder. This miniaturization presents considerable engineering hurdles.

High-Pressure Requirements

Delivering liquid hydrogen at the needed high-pressure levels for injection can cause the internal pump to fail owing to frictional heat. This requires robust pump designs and cooling systems to manage the high-pressure hydrogen delivery (Dash et al. 2022).

Safety Concerns

The hydrogen is highly flammable and has wide range of flammability limits. These properties of hydrogen are making it unsafe in case of leaks. Hydrogen is lighter and has higher diffusivity which make it safer in open environment.

4.7 Conclusions

Steam reforming, partial oxidation of hydrocarbons, and coal gasification are all ways for producing hydrogen, each with their own set of advantages and challenges. Steam reforming, while commonly utilized, produces large carbon dioxide emissions because to its dependency on fossil fuels. Although partial oxidation allows for the use of multiple fuel sources, it also produces CO and CO_2. Coal gasification generates syngas but emits hazardous pollutants including sulphur dioxide and nitrogen oxides. Green hydrogen generation from renewable sources like solar, wind, and hydro power offers a more sustainable option by utilizing electrolysis, which separates water into hydrogen and oxygen using electricity from renewable sources.

Hydrogen is a possible alternative fuel for internal combustion engines, with considerable environmental and performance advantages. The trials show that hydrogen may effectively substitute diesel, increasing efficiency while decreasing specific energy consumption (SEC). For example, with a hydrogen flow rate of 7.5 l/min, the brake thermal efficiency (BTE) was 25.6%, versus 21.6% for diesel at 75% load. This boost in efficiency is accompanied by a significant fall in SEC, demonstrating hydrogen's potential to improve engine performance while lowering fuel consumption.

Furthermore, hydrogen combustion produces reduced levels of CO and smoke, solving some of the key environmental concerns connected with traditional fuels. However, the use of hydrogen causes a modest increase in hydrocarbon (HC) emissions and poses issues in terms of NOx emissions, which tend to rise at greater flow rates due to the higher combustion temperatures of hydrogen.

In terms of combustion properties, hydrogen allows for higher peak heat release rates and more efficient combustion cycles. The peak pressures achieved with hydrogen are comparable to those of diesel, and the speedier combustion process improves overall engine efficiency.

The use of hydrogen in internal combustion engines raises various issues, including backfire and knock, storage and handling, material compatibility, and durability. Backfire is caused by hydrogen's high rate of combustion and low ignition energy, which leads to premature ignition. Because of hydrogen's low energy density, it must be stored and handled under high pressures. The high diffusivity of hydrogen causes material compatibility concerns, which can lead to engine component embrittlement. Furthermore, the lack of lubricate qualities of hydrogen needs specialized injector designs to reduce wear and tear. NOx emissions at high combustion temperatures are also a concern, necessitating advanced techniques to reduce their formation. Despite these challenges, hydrogen remains an intriguing alternative fuel since it has the ability to cut greenhouse gas emissions and other pollutants.

Overall, the use of hydrogen in ICEs has enormous promise for lowering reliance on fossil fuels and mitigating environmental damage. However, improving hydrogen flow rates and managing emission issues are important for realizing the full promise of this alternative fuel.

References

Acar C, Dincer I (2014) Comparative assessment of hydrogen production methods from renewable and non-renewable sources. https://doi.org/10.1016/j.ijhydene.2013.10.060

Ajeeb W, Costa Neto R, Baptista P (2024) Life cycle assessment of green hydrogen production through electrolysis: a literature review. Elsevier Ltd. https://doi.org/10.1016/j.seta.2024.103923

Benghanem M et al (2023) Hydrogen production methods based on solar and wind energy: a review. MDPI. https://doi.org/10.3390/en16020757

Chao Y, Huang CT, Lee HM, Chang MB (2008) Hydrogen production via partial oxidation of methane with plasma-assisted catalysis. Int J Hydrogen Energy 33(2):664–671. https://doi.org/10.1016/j.ijhydene.2007.09.024

Dash SK, Chakraborty S, Roccotelli M, Sahu UK (2022) Hydrogen fuel for future mobility: challenges and future aspects. https://doi.org/10.3390/su14148285

Dash SK, Chakraborty S, Elangovan D (2023) A brief review of hydrogen production methods and their challenges. MDPI. https://doi.org/10.3390/en16031141

Executive summary—CO_2 emissions in 2023—analysis—IEA. Accessed on 17 Dec 2024 [Online]. Available: https://www.iea.org/reports/co2-emissions-in-2023/executive-summary

Edwards PP, Kuznetsov VL, David WIF (2007) Hydrogen energy. Philos Trans Roy Soc Math Phys Eng Sci 365(1853):1043–1056. https://doi.org/10.1098/rsta.2006.1965

FCEVs to be cheaper to run than BEVs and ICE vehicles within 10 years | Deloitte China | Press release. Accessed on 17 Dec 2024. Available: https://www2.deloitte.com/cn/en/pages/about-deloitte/articles/pr-fcevs-to-be-cheaper-to-run-than-bevs-and-ice-vehicles-within-10-years.html

Faizal M, Chuah LS, Lee C, Hameed A, Lee J, Shankar M (2019) Review of hydrogen fuel for internal combustion engines. J Mech Eng Res Dev 42(3):36–46. https://doi.org/10.26480/jmerd.03.2019.35.46

Gao J, Wang X, Song P, Tian G, Ma C (2022) Review of the backfire occurrences and control strategies for port hydrogen injection internal combustion engines. Elsevier Ltd. https://doi.org/10.1016/j.fuel.2021.121553

Genç MS, Çelik M, Karasu I (2012) A review on wind energy and wind-hydrogen production in Turkey: a case study of hydrogen production via electrolysis system supplied by wind energy conversion system in Central Anatolian Turkey. https://doi.org/10.1016/j.rser.2012.08.011

Gomes Antunes JM, Mikalsen R, Roskilly AP (2009) An experimental study of a direct injection compression ignition hydrogen engine. Int J Hydrogen Energy 34(15):6516–6522. https://doi.org/10.1016/j.ijhydene.2009.05.142

Gupta RB (2008) Hydrogen fuel production, transport, and storage

H production from coal hydrogen fact sheet series facts [Online]. Available: www.HydrogenAssociation.org

Hosseini SE, Butler B (2020) An overview of development and challenges in hydrogen powered vehicles. Taylor and Francis Inc. https://doi.org/10.1080/15435075.2019.1685999

Hosseini SH et al (2023) Use of hydrogen in dual-fuel diesel engines. Elsevier Ltd. https://doi.org/10.1016/j.pecs.2023.101100

Hwang HT, Varma A (2014) Hydrogen storage for fuel cell vehicles. Elsevier Ltd. https://doi.org/10.1016/j.coche.2014.04.004

Ishaq H, Dincer I (2021) Comparative assessment of renewable energy-based hydrogen production methods. Renew Sustain Energy Rev 135. https://doi.org/10.1016/j.rser.2020.110192

Ishaq H, Dincer I, Crawford C (2022) A review on hydrogen production and utilization: challenges and opportunities. Int J Hydrogen Energy 47(62):26238–26264. https://doi.org/10.1016/j.ijhydene.2021.11.149

Jayaprabakar J et al (2024) Prospectus of hydrogen enrichment in internal combustion engines: methodological insights on its production, injection, properties, performance and emissions. Fuel 363. https://doi.org/10.1016/j.fuel.2024.131034

Ji M, Wang J (2021) Review and comparison of various hydrogen production methods based on costs and life cycle impact assessment indicators. Elsevier Ltd. https://doi.org/10.1016/j.ijhydene.2021.09.142

Kamble AD, Saxena VK, Chavan PD, Mendhe VA (2019) Co-gasification of coal and biomass an emerging clean energy technology: Status and prospects of development in Indian context. China University of Mining and Technology. https://doi.org/10.1016/j.ijmst.2018.03.011

Kamboj P, Malyan A, Kaur H, Jain H, Chaturvedi V (2022) India transport energy outlook. Council on Energy, Environment and Water, New Delhi

Kothari R, Buddhi D, Sawhney RL (2008) Comparison of environmental and economic aspects of various hydrogen production methods. https://doi.org/10.1016/j.rser.2006.07.012

Kumar N, Shojaee M, Spivey JJ (2015) Catalytic bi-reforming of methane: from greenhouse gases to syngas. Elsevier Ltd. https://doi.org/10.1016/j.coche.2015.07.003

de Medeiros FGM, Lopes FWB, Rego de Vasconcelos (2022) Prospects and technical challenges in hydrogen production through dry reforming of methane. MDPI. https://doi.org/10.3390/catal12040363

Onorati A et al (2022) The role of hydrogen for future internal combustion engines. SAGE Publications Ltd. https://doi.org/10.1177/14680874221081947

Path to hydrogen competitiveness: a cost perspective | hydrogen council. Accessed on 17 Dec 2024 [Online]. Available: https://hydrogencouncil.com/en/path-to-hydrogen-competitiveness-a-cost-perspective/

Rasul MG, Hazrat MA, Sattar MA, Jahirul MI, Shearer MJ (202) The future of hydrogen: challenges on production, storage and applications. Elsevier Ltd. https://doi.org/10.1016/j.enconman.2022.116326

Saha P et al (2024) Grey, blue, and green hydrogen: a comprehensive review of production methods and prospects for zero-emission energy. Taylor and Francis Ltd. https://doi.org/10.1080/15435075.2023.2244583

Saravanan N, Nagarajan G, Dhanasekaran C, Kalaiselvan KM (2007) Experimental investigation of hydrogen fuel injection in DI dual fuel diesel engine

Shadidi B, Najafi G, Yusaf T (2021) A review of hydrogen as a fuel in internal combustion engines. MDPI. https://doi.org/10.3390/en14196209

Stępień Z (2021) A comprehensive overview of hydrogen-fueled internal combustion engines: achievements and future challenges. MDPI. https://doi.org/10.3390/en14206504

Sun Z et al (2023) Hydrogen engine operation strategies: recent progress, industrialization challenges, and perspectives. Elsevier Ltd. https://doi.org/10.1016/j.ijhydene.2022.09.256

Toyota Mirai | Toyota.com (2024). Accessed on 17 Dec 2024 [Online]. Available: https://www.toyota.com/mirai/

Taji M, Farsi M, Keshavarz P (2018) Real time optimization of steam reforming of methane in an industrial hydrogen plant. Int J Hydrogen Energy 43(29):13110–13121. https://doi.org/10.1016/j.ijhydene.2018.05.094

Tongur S (2018) Preparing for takeoff analyzing the development of electric road systems from a business model perspective

Ture E (2007) Hydrogen production from solar energy. In: Sheffield John Ç, Sheffield W (ed) Assessment of hydrogen energy for sustainable development. Springer Netherlands, Dordrecht, pp 135–146

Use of hydrogen in internal combustion engine [Online]. Available: www.erpublication.org

Wozniak M, Ozuna G, Siczek K (2021) Problems with glow plug—a review. Combust Eng 186(3):11–30. https://doi.org/10.19206/ce-140114

Zohuri B (2018) Hydrogen energy: challenges and solutions for a cleaner future. Springer International Publishing. https://doi.org/10.1007/978-3-319-93461-7

Chapter 5
Role of Ammonia as a Sustainable Fuel: Production Methods, Properties, Safety, and Applications in Fuel Cells and IC Engines

Ankit Debnath, Krishna Kumar Pandey, Paramvir Singh, and Sudarshan Kumar

Abstract Ammonia has existing infrastructure, is carbon-free, has high gravimetric and volumetric hydrogen densities, and can be used directly without breaking down, making it a suitable hydrogen storage material. Ammonia is also considered safe because of its high autoignition temperature, low condensation pressure, and lower gas density than air. Ammonia may be generated from a wide range of primary energy sources, such as fossil fuels, renewable energy, and excess energy. Ammonia energy may be directly extracted and used as fuel at the utilization site. Ammonia is produced using the electrochemical, thermochemical, and Haber–Bosch cycle methods that are now in use. Adopting ammonia as fuel is encouraging, but there are several obstacles which still need to be overcome for wider and more effective combustion. Since ammonia has a moderate flame speed and a high ignition energy, it is not very flammable. The study highlights ammonia's potential for direct use in internal combustion engines and direct ammonia fuel cells due to its superior overall energy efficiency. However, utilization of ammonia is not difficult as it may result in NO_x emissions and poor combustion performance. Ammonia may assist in decarbonizing current ammonia markets by replacing fossil fuels and promoting greening in the industrial sector, such as fertilizer and chemical manufacturing. Ammonia can also accelerate decarbonization in businesses that are difficult to abate, such as by retrofitting already-existing ammonia facilities.

Keyword Ammonia · Electrochemical synthesis · Fuel cell · Internal combustion engines

---

A. Debnath · K. K. Pandey · P. Singh (✉)
Department of Mechanical Engineering, National Institute of Technology Agartala, Agartala, Tripura, India
e-mail: psingh.me@nita.ac.in

S. Kumar
Department of Aerospace Engineering, Indian Institute of Technology Bombay, Mumbai, India

N. Sharma et al. (eds.), *Sustainable, Scalable and Storable E-Fuels for Decarbonising Transport Sector*, Energy, Environment, and Sustainability,
https://doi.org/10.1007/978-981-96-7388-9_5

5.1 Introduction

Fuel for internal combustion engines (ICEs) has to have more sophisticated qualities than only the capacity to burn. Because ICEs require a fuel that can be used under various operating situations (loads and speeds), the current preferences for gasoline and diesel are not correlated. These fuels fulfill the superior qualities in a variety of application domains. Among the characteristics are flame and ignition spread. There are certain benefits and drawbacks of using ammonia in ICE (Sun et al. 2024). Ammonia offers the advantage of being a carbon-free fuel, meaning it does not release CO_2 or produce soot when burned. However, despite these benefits, ammonia also has certain drawbacks, such as a lower flame speed, which pose challenges for combustion in ICEs. Using ammonia fuel can lead to low cost of petroleum, which is not much lead done on the subject (Reiter and Kong 2011).

Ammonia possesses high minimum ignition energy and high ignition temperature, along with narrow flammability limits and lower flame propagation speed (Zhang et al. 2023). When these four characteristics come together, achieving appropriate combustion at different loads and speeds can be challenging. Moreover, ammonia vaporizes at high heat, causing a large temperature to decrease when, as ammonia turns from a liquid to a gas, the temperature within the cylinder drops. Additionally, issues with it are possible for injection nozzles to freeze. However, by entering as fuel through the intake manifold, the ammonia vapor displaces an equal quantity of air, resulting in a fall in volumetric efficiency (Kumar et al. 2024). Conversely, the significant temperature drop during combustion can be prevented by premixing the fuel and air in a gaseous form. However, the nitrogen content in ammonia still poses a challenge, as it contributes to NO_x emissions (Kumar et al. 2024).

5.1.1 Needs of Ammonia

Hydrogen and ammonia can be considered the most important and clean fuel source with zero-carbon emission, and it also increases trend for its use in the future (Withagen 1994; Zamfirescu and Dincer 2008). Ammonia's unique properties and environmental benefits make it an ideal alternative to conventional fossil fuels. Additionally, ammonia can store hydrogen, which is three times more expensive than ammonia. In the hydrogen economy, where ammonia can serve both as a means of energy storage and transportation, it will be highly beneficial for the future (Tawalbeh et al. 2022). One of the key advantages of ammonia is its zero-carbon emissions during combustion. Additionally, when combined with renewable energy sources, ammonia can be used as a clean fuel and produced sustainably, making it a highly environmentally friendly energy source.

5.1.2 *Use of Ammonia*

Ammonia has been considered as a fuel since the early nineteenth century, reflecting a long history of its utilization. Interest in using ammonia as a fuel has occasionally resurfaced, especially during energy crises and growing environmental concerns. Ammonia presents an interesting backup for fossil fuels because of its high hydrogen concentration, which can result in lower greenhouse gas emissions and carbon-free burning. Ammonia has become increasingly popular as a sustainable fuel due to its unique features, quality, and capacity to express a few energy and environmental problems (Kumar and Sleiti 2024).

Ammonia burns cleanly and produces just nitrogen and water vapor, which comparably make less greenhouse gas emissions when compared to conventional fossil fuels. This is one of ammonia's key interests. Moreover, it has a greater energy density than hydrogen, making transportation and storage quite easier. Ammonia is easier to unite into the current fuel infrastructure and architecture since it can be liquefied at cryogenic temperatures or at relatively low pressures (MacFarlane et al. 2020). Ammonia can also be used for a variety of other purposes, such as power generation, transportation, and the delivery and capacity of hydrogen for energy distribution. Considering ammonia as a sustainable fuel, there may be a significant step in addressing climate change and reducing our need for fossil fuels in the direction of a low-carbon future. For ammonia to become a significant part of the future energy landscape, it must overcome challenges such as improving production methods, updating infrastructure, and addressing safety issues related to its toxicity and corrosiveness. (Adeli et al. 2023).

5.2 Production Methods

There are specific production methods through which it can be used as a sustainable fuel.

5.2.1 *Haber–Bosch Process*

The most common technique for producing ammonia is the Haber–Bosch process, which involves a catalyst-assisted reaction between ambient nitrogen and hydrogen at high pressure and temperature. The conventional method of producing hydrogen from natural gas (methane) involves steam methane reforming (SMR), which emits CO_2. Using renewable energy sources, hydrogen may be produced by water electrolysis to produce ammonia in a more environmentally friendly manner (Ghavam et al. 2021). With the proper adjustments, ammonia generated by the Haber–Bosch process may

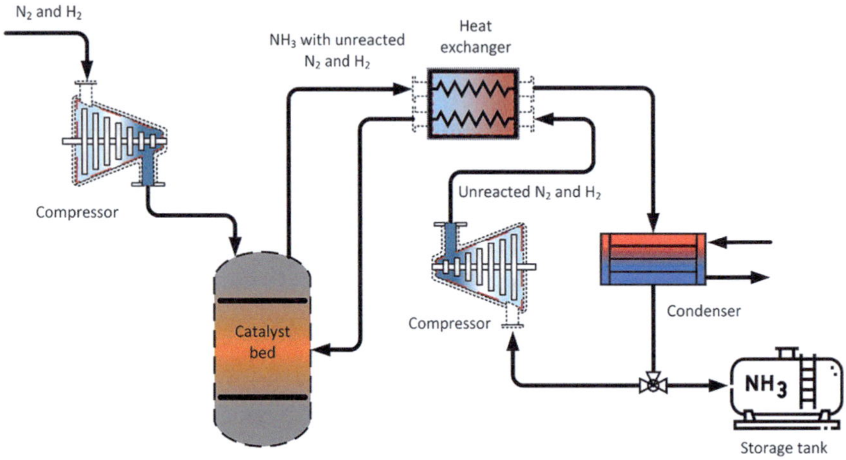

Fig. 5.1 Simple layout of Haber–Bosch ammonia synthesis process (Ishaq and Crawford 2024)

be kept as a liquid at relatively low pressure and temperature, making it suitable for fuel in ICEs.

However, most of the time, this process is used in high temperatures and high pressure to produce ammonia, and renewable sources can also use it. The overall efficiency is 60–70, and 80% is the electrolyte efficiency; all the system is efficiency depends on renewable energy (Ishaq and Crawford 2024). The thermos-catalytic reaction is formed of the Haber–Bosch is shown in Eq. (5.1) (Yapicioglu and Dincer 2019) (Fig. 5.1),

$$\frac{3}{2}H_2 + \frac{1}{2}N_2 \rightarrow \ NH_3 + 45.2\,(kJ/Mole) \tag{5.1}$$

5.2.2 *Thermocatalytic Ammonia Synthesis*

Since ammonia is needed for fertilizers and other chemicals, thermocatalysis, primarily achieved via the Haber–Bosch process, is crucial to modern agriculture and industry. In this process, nitrogen and hydrogen gasses are reacted at high pressures (150–300 atm) and temperatures (400–500 °C) over an iron-based catalyst (Araia et al. 2022). For durability and efficiency, the catalyst which is frequently improved with additions like potassium and aluminum oxide is crucial. Nitrogen's strong triple bond makes it difficult to dissociate. Hence, catalyst selection and working conditions are crucial (Yang et al. 2023). The average high temperature in this synthesis is 390–510 °C, and the high pressure is from 140 to 310 bars. To break the nitrogen triple bond, the catalyst plays an essential role to low the activation energy. Nowadays,

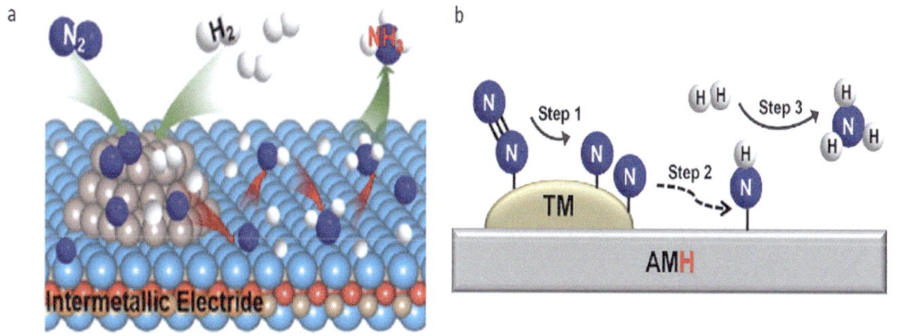

Fig. 5.2 Thermocatalytic ammonia synthesis (Verschoor et al. 2024)

advanced research has developed on increasing the efficiency, energy consumption and explores alternative catalysts, low preserve method to make this process more sustainable. Thermocatalyst system has an impactful on good environmental implementation, which includes the emission of greenhouse gases (Fig. 5.2).

5.2.3 Liquid Electrolyte-Based Systems

Ethanol provides the hydrogen needed for liquid electrolyte-based ammonia production systems, while lithium perchlorate in tetrahydrofuran acts as the electrolyte. The efficiency might theoretically be raised by altering the temperature and pressure settings to the optimal levels for manufacturing. There may have been some long-term problems with this approach because ionic liquid electrolytes were also shown to be present during the trials. An additional experiment producing ammonia electrochemically in a molten electrolyte utilizing $CoFeO_x$ and nano-Fe_2O_3 as catalysts demonstrated enhanced nitrogen reduction efficiency, suggesting the potential of mixed metal oxide catalysts in high-temperature molten salt systems for sustainable ammonia synthesis (Yapicioglu and Dincer 2019). Rechargeable batteries are one of the common applications of liquid electrolyte-based systems, such as lithium-ion batteries. The electrolyte in these batteries allows the movement of lithium ions between the positive (anode) to negative (cathode) electrodes during the charging and discharging time. However, the battery performance, including conductivity, stability, and electrochemical compatibility in the electrode, can be tough and challenging if choose a liquid electrolyte. For this reason, organic solvents such as ethylene carbonate and dimethyl carbonate are used in lithium-ion batteries (Liao et al. 2020) (Fig. 5.3).

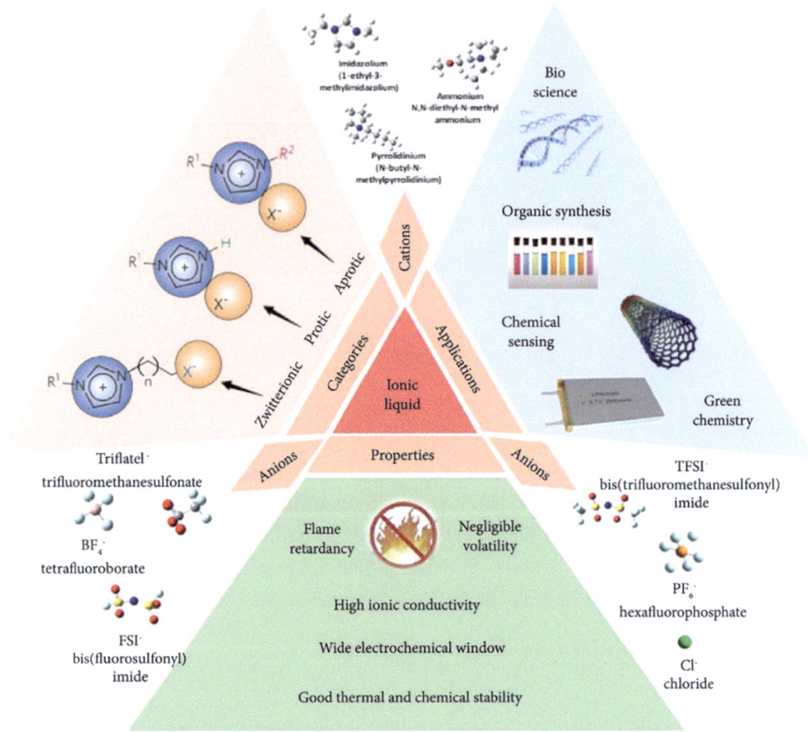

Fig. 5.3 Liquid electrolyte-based systems (Zhu et al. 2021)

5.2.4 Composite Membrane-Based System

The composite electrolytes are constructed by assembling different ionic conducting phases, and further phases are added to the primary phases to determine the thermal, mechanical, and electrical characteristics of the ammonia that is created. Under experimental settings, it has been demonstrated that an oxide and an alkali metal carbonate exhibit proton, carbonate, and oxygen-ion conductivity. According to research on possible electrolytes for fuel cells operating at moderate temperatures (400–800 °C) focuses on composite membrane-based systems. These electrolytes are tested under various operating settings (Yapicioglu and Dincer 2019). The advantage of this composite membrane-based system is that it has the tendency to overcome all the limitations of single-layer membranes, such as gas separation. It can achieve high sensitivity by maintaining reasonable permeability, which may be dangerous for the operation. In the other case, water desalination, the rejection ratio of salt is high, and other products allow a high flow of water (Golchoobi et al. 2020). In case of energy storage and conversion system, suppose a fuel cell battery can play an important role as electrolyte for improving the device's efficiency, durability, and safety (Fig. 5.4).

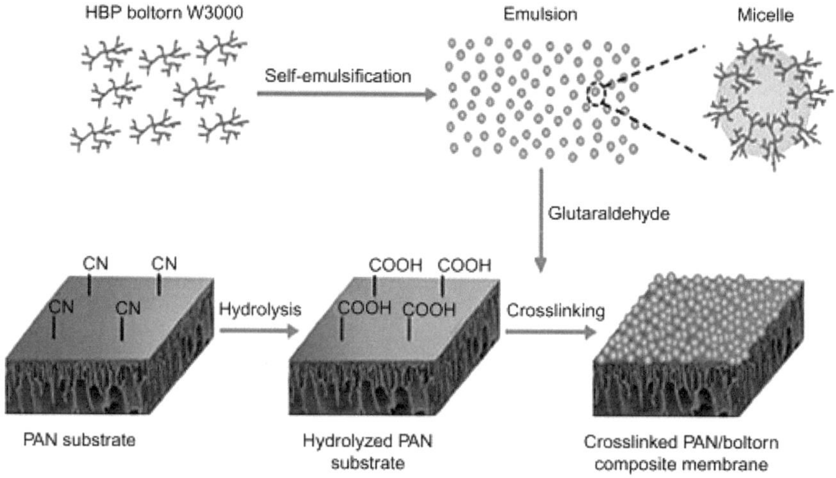

Fig. 5.4 Composite membrane-based system (Wang et al. 2014)

5.2.5 *Solid-State Electrolyte-Based System*

The choice of proton and mixed proton/oxygen-ion conducting solid systems is used to test the electrolytes in the research on electrochemical ammonia production (Giddey et al. 2013). In these, systems are anodes and cathodes with a compact solid electrolyte that acts as a barrier to gas diffusion and allows protons or oxide ions to transfer between them. Solid-state proton conductors are a family of ionic solid electrolytes that have the ability to transmit and use hydrogen ions directly (Giddey et al. 2013). However, the approach is less feasible due to the requirement for subsequent phases and the unavoidably high working temperatures (Yapicioglu and Dincer 2019). One of the heavy advantages of this electrolyte is that it has the potential to increase the density of energy in batteries. It has limitations in the properties of liquid electrolytes. Solid-state batteries can use a lithium metal anode; it will allow much high-capacity energy. This will make batteries life span increase, more storage capacity, and will make attractive for EV's and renewable energy systems. Figure 5.5 shows the production and utilization of ammonia.

According to this picture, at first there are different sources from which resources of production material are collected or synthesized. So, the different sources are wind, solar, geothermal, biomass, nuclear, and excessive electricity. So, these are different sources from which production has been done. The production process starts with the help of electricity and high-temperature heat supply the separation of hydrogen and oxygen is done. The production of hydrogen with the help of various production processes like the Haber–Bosch process, electrochemical synthesis, plasma-assisted synthesis, thermocatalytic ammonia synthesis, liquid electrolyte-based systems, composite membrane-based systems, and solid-state electrolyte-based systems reflects a diverse and evolving technological landscape

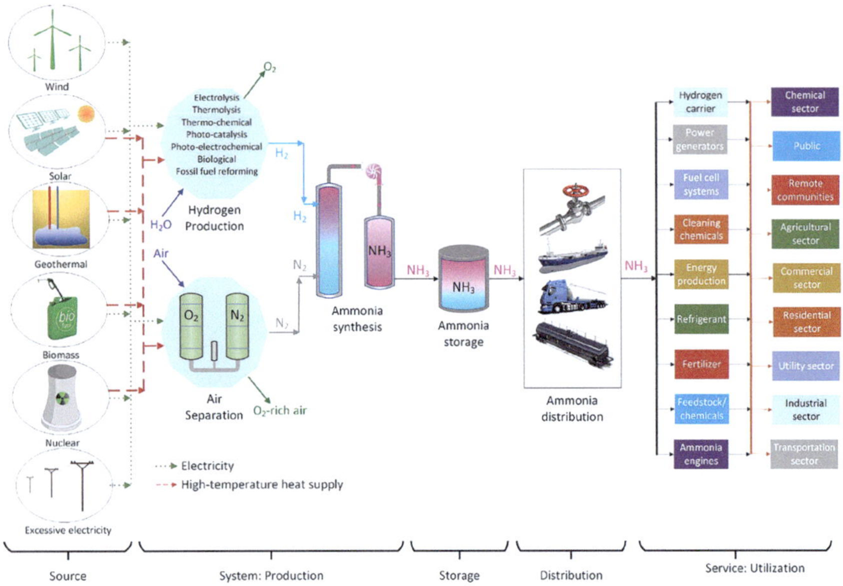

Fig. 5.5 Economic cycle of ammonia from production to utilization (Erdemir and Dincer 2021)

aimed at improving efficiency, reducing carbon emissions, and enabling decentralized hydrogen generation. The air has been separated and the oxygen and nitrogen are stored. The production of ammonia has proceeded. Here from the stored containers, hydrogen and nitrogen are mixed and NH_3 is produced and contained in a long and stored vessel. After storage, the important one is distribution. The distribution has been done through various processes like pipeline and marine transport, by road transport and by ammonia vessels. For successful transportation, ammonia serves various functions, including as a hydrogen carrier, in the chemical sector, power generation, fuel cell systems, remote communities, cleaning chemicals, agriculture, energy production, commercial sectors, refrigeration, residential utilities, feedstock, industrial applications, ammonia engines, and the transportation sector.

5.2.6 Electrochemical Synthesis

This more modern and eco-friendly method employs energy, sometimes from renewable sources, to directly produce ammonia from nitrogen and water through an electrochemical process. Given that it does not need hydrogen production from natural gas, it may be more efficient and ecologically benign. Ammonia produced by electrochemical synthesis can be used directly as fuel in ICEs when zero-carbon fuels are desired. This process is also known as electrochemical reduction because it reduces the electric power of H_2O and N_2 to produce NH_3. This process drives the reaction

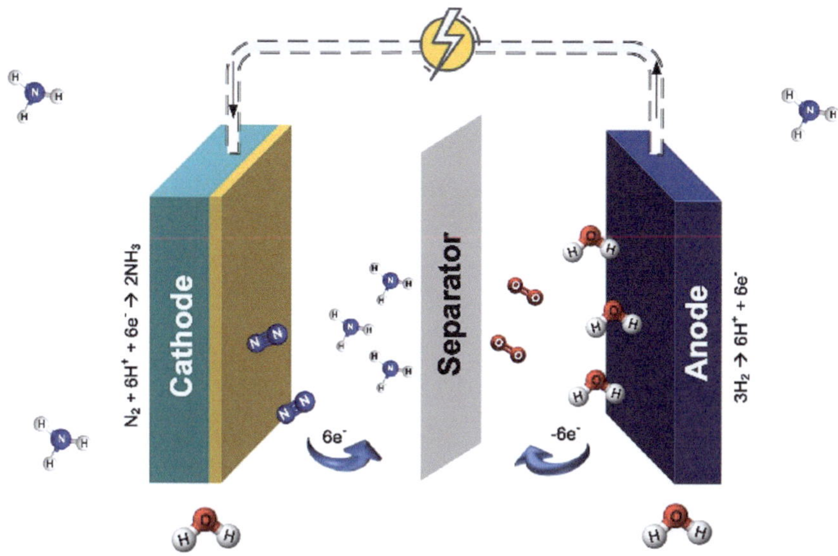

Fig. 5.6 Basic illustration of electrochemical ammonia synthesis process (Ishaq and Crawford 2024)

of N_2 and H_2 for the ammonia synthesis (Shen et al. 2021). For this process, the most important thing is to find the perfect catalyst analyze the reaction at low pressure and temperature (Shen et al. 2021; Montoya et al. 2015) (Fig. 5.6).

5.2.7 *Plasma-Assisted Synthesis*

Unlike the Haber–Bosch process, plasma-assisted synthesis activates nitrogen and hydrogen using an ionized gas called plasma, enabling the combination to produce ammonia at lower pressures and temperatures (Faria 2021). This method can lead to a reduction in energy consumption and is suitable for smaller-scale, decentralized production. Ammonia produced via plasma-assisted synthesis can be used in ICEs, particularly in situations where small production units are advantageous. In this process, H_2 and N_2 are exposed to plasma, which help break the nitrogen bonds, which is a critical step in ammonia production. Its efficiency depends upon some various factors like plasma reduction design, operational condition, and presence of a catalyst. However, this technique presents a challenge because the process generates a significant amount of heat, and the high temperature can cause ammonia to decompose into its constituents (Ishaq and Crawford 2024) (Fig. 5.7).

In conventional ammonia production, CO_2 emissions arise from two primary sources, fossil fuel combustion and process emissions. During fossil fuel combustion, hydrogen required for ammonia synthesis is obtained from natural gas through

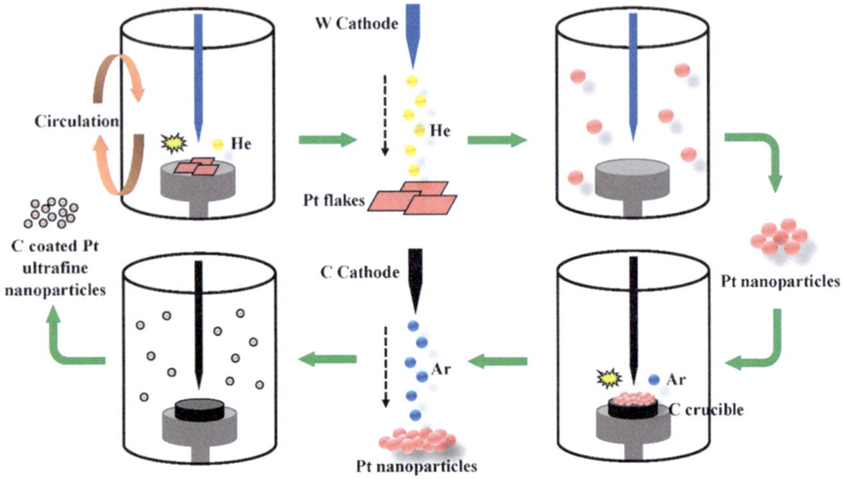

Fig. 5.7 Plasma-assisted synthesis diagram (Zhang et al. 2020)

steam reforming. This process generates CO_2 as a byproduct when steam reacts with methane to produce hydrogen and carbon monoxide. Additionally, process emissions contribute to the overall carbon footprint, further exacerbating environmental concerns. While conventional ammonia production is highly efficient, its dependence on fossil fuels renders it environmentally unsustainable, leading to significant greenhouse gas emissions and emphasizing the need for cleaner production alternatives.

The transition to green ammonia production is crucial for mitigating the environmental impact of ammonia synthesis. Green methods focus on integrating renewable energy sources and employing innovative technologies to reduce carbon emissions. Among these, electrolysis-based production is currently the most advanced and commercially viable. Additionally, emerging technologies such as photocatalysis and plasma-assisted synthesis hold great potential for achieving sustainable ammonia production. The need for green ammonia production is increasingly recognized as critical for addressing global environmental challenges. This shift is driven by the necessity to decarbonize ammonia production, combat climate change, facilitate the energy transition in agriculture and industry, promote sustainable global practices, and meet growing economic and regulatory demands.

5.3 Properties of Ammonia

Ammonia must address challenges such as developing efficient and cost-effective production methods and modifying infrastructure to fully realize its potential as a key player in the future energy landscape. Additionally, ammonia presents a more

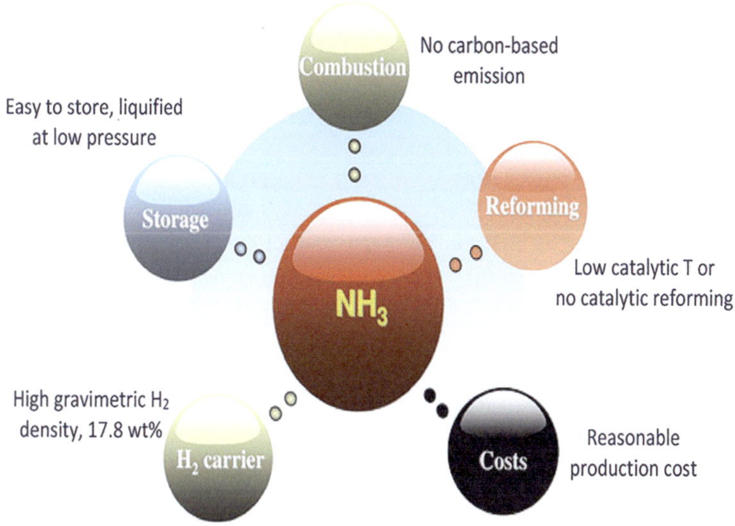

Fig. 5.8 Properties of ammonia as a fuel (Dimitriou and Javaid 2020)

sustainable option compared to biofuels. Biofuels frequently depend on agricultural feedstocks, which can conflict with food production and result in environmental issues such as deforestation (Ben-Iwo et al. 2016). The density of ammonia energy is lower than that of traditional fuels like gasoline or diesel. Its energy content is approximately 50% less than that of gasoline. Compared to other fuels, ammonia is harder to ignite and maintain combustion due to its comparatively high ignition temperature. Ammonia has a lower flame speed, which causes slower combustion and necessitates engine designs that are designed for efficient use. Figure 5.8 depicts the general properties of ammonia. Table 5.1 illustrates the chemical properties of ammonia.

5.4 Storage, Transportation, and Safety Concerns of Ammonia

Ammonia is relatively easy to store and transport because it remains liquid at low temperatures and moderate pressures. However, ammonia is corrosive and toxic, making leaks potentially hazardous, so proper handling and storage procedures are essential. Although ammonia is flammable, it has a limited flammability range in the air (approximately 15–25% by volume), making it less flammable compared to other fuels.

Table 5.1 Comparison of fuel properties (Chong and Hochgreb 2011; Eyidogan et al. 2010)

	Ammonia (liquified)	Hydrogen (liquified)	Diesel	Gasoline	Methanol	Ethanol
Energy content (LHV) (MJ/kg)	18.6	120	44.11	44.34	19.9	26.84
Density (kg/m^3)	626	70.85	745.7	697.6 (n-octane)	786.3	785.1
Octane (RON)	>130	>130	<20	100	108.7	108.6
Flame velocity (m/s)	0.067	3.25	~0.8	0.41–0.58	0.56	0.58
Flammability limits (vol/%)	15–28	4.5–75	0.43–0.6	0.95–6	6.7–36	3.3–19
Minimum ignition energy (mJ)	680	~0.016	~0.23	1.35	10.14	0.65

5.4.1 *Storage*

The storage and transportation of ammonia are comparatively more practical and costeffective than hydrogen. Besides hydrogen, which needs cryogenic or high-pressure storage, ammonia might be stored as a liquid under moderate or low pressure, making it more practical, genuine, and cost-effective. Because ammonia is heavily used in the fertilizer industry, more infrastructure for production and distribution can be used for fuel. Ammonia storage depends on how much we can convert to renewable energy. So, the process of renewable energy storage is done by converting electricity into chemical energy stored in ammonia (Ishaq and Crawford 2024). Ammonia can be stored in a liquid state in large amounts with perspective to different pressure changes from 10 to 15 bars and cooled up to −33 °C, making ammonia a better and good potential renewable energy storage.

Ammonia storage as a liquid can be done using another method like propane (Aziz et al. 2020). It can be done in two processes; one is by increasing pressure, and another is by decreasing temperature. So, when pressure is increasing the temperature must be maintained at 25 °C, while in the second case, decreasing the temperature, maintain the pressure at atmospheric level. Therefore, the economic and overall performance of the ammonia storage will be countable and well engaged and will be established better.

5.4.2 *Transportation*

Ammonia transportation method becomes one of the crucial aspects in ammonia production. So, the ammonia transportation method has various factors to improve the quality of ammonia to be transported; (i) there is a possibility in where we can use pipelines which is very cost-effective and more efficiently ammonia can be transported, and though it can be transported long distance, (ii) the use of road tankers by which ammonia can be transported through a medium, short distance and it has designed to deliver it safely; (iii) the rail transport management and it comes under the most common transportation technique over short medium distance; and (iv) the use of ammonia vessels through which huge quantity of ammonia can be transported.

The boiling points of ammonia and propane are −33.34 and −42.1 °C, respectively. Both substances exhibit the same condensation pressure at these boiling points of 1 atm. However, transport ships designed for propane cannot be used for ammonia without modifications. The necessary changes include:

(i) Material compatibility: Ammonia is highly corrosive to carbon steel, copper, and copper alloys. To prevent damage, tanks, pipelines, and valves must be constructed from stainless steel or coated with ammonia-compatible materials.
(ii) Safety enhancements: Ammonia is toxic and poses serious health risks if leaked. This requires the installation of advanced safety systems, such as gas detection devices and emergency response protocols, to mitigate hazards. Due to its unique chemical properties, these modifications are essential for safely transporting ammonia.

5.4.3 Safety Concern

Nowadays, ammonia is very important and plays a vital role in global energy transition, but sometimes mistakenly, some accidents occur, which can be minor, but this minor may be more effective and cause danger. It can cause damage and have left some serious impacts on environments and other places accordingly. As we know, ammonia has some toxic effects that can damage biodiversity, such a slants, animals, and aquatic organisms (Edwards et al. 2024). Ammonia used in current engines is not environmentally friendly because these engines are not properly modified or optimized for ammonia combustion, leading to incomplete or inefficient burning. This incomplete combustion can result in the emission of harmful gases, including nitrogen oxides (NO_x), which are potent pollutants that pose health risks and contribute to atmospheric and climate changes. So, some safety concerns and challenges may need to be solved by taking steps (Ishaq and Crawford 2024).

(a) The most important is that the place where ammonia is stored should have proper ventilation across all around; otherwise, it may affect harmful danger.
(b) If there is any leak, it may cause a problem; a leak detection system should be installed to detect leakages at very low concentrations.
(c) There should be regular inspection and maintenance in all ammonia related objects, avoid errors while doing work and get rid from damage and corrosion.
(d) One of the important things is that which we need to check that proper design and installation have done or not because the system and projects involve with

ammonia handling, storage, and utilization, that should meet safety standards and regulations.

5.5 Application of Ammonia in Fuel Cell

Ammonia may be utilized to create power in fuel cells. It can break down into hydrogen and nitrogen, with the hydrogen subsequently being utilized in typical hydrogen fuel cells, or it can be fed straight into specialist ammonia fuel cells. In fuel cells, the process occurs some chemical reactions shown in Eqs. (5.2–5.5):

Anodic Reaction:

$$2NH_3 \rightarrow N_2 + 3H_2 \tag{5.2}$$

$$3H_2 \rightarrow 6H^+ + 6e^- \tag{5.3}$$

Cathodic Reaction:

$$3O_2 + 6e^- \rightarrow 3O_2^- \tag{5.4}$$

$$6H^+ + 3O_2^- \rightarrow 3H_2O \tag{5.5}$$

Basically, fuel cell is also known as technology in that can generate electricity from the chemical resources based on electrochemical reaction (Mekhilef et al. 2012). Fuel cell is categorized into several types depending on their electrolyte material, function application, and operating condition (Minh 1993). They are;

1. Alkaline membrane fuel cell (AMFC).
2. Direct methanol fuel cell (DMFC).
3. Molten carbonate fuel cell (MCFC).
4. Phosphoric acid fuel cell (PAFC).
5. Alkaline fuel cell (AFC).
6. Solid oxide fuel cell (SOFC).
7. Proton exchange membrane fuel cell (PEMFC).

So, among all of them, SOFC is taken as a most promising one because of its high energy intensity, efficiency, high fuel flexibility, and environmentally friendly. The operating temperature is between 600–1000 °C.

(i) AFC: The first cell which developed to run with ammonia is alkaline fuel cell. It runs at low temperature from 50 to 200 °C. It uses potassium hydroxide (KOH) electrolyte and molten hydroxide (NaOH + KOH) electrolyte, to operate at higher temperature.

(ii) AMFC: In alkaline membrane fuel cell, there are two modes based on which the process works; one is it has anode function for ammonia, and another is cathode function for oxygen-containing gas and this operation has been done at a temperature of 80 °C.

(iii) PAFC: This fuel cell uses a liquid phosphoric to convert chemical energy into electrical energy. It was developed in the mid-1960–1970. The operating temperature will be from 150 to 200 °C. The electrode that is being used in that process is a platinum catalyst.

(iv) MCFC: The molten carbonate fuel cell has been designed and created to achieve small to large energy distribution power on generating system and the process output will be 0.3–3 mw range. The operating temperature is between 600 and 700 °C.

(v) PEMFC: In this case, a membrane converts hydrogen and oxygen into electricity. The membrane is made with semipermeable polymer. It was made to use its function to provide facility to transport, stationary, and small adjustable application.

(vi) SOFC: This device converts chemical energy of fuel directly to electrical energy, at high temperatures. The operating temperature is between 600 to 1000 °C.

(vii) DMFC: It comes under the category of PEMFC, in that rather than the use of hydrogen methanol is directly feed into the fuel cell. And it has low energy density and efficiency. It has also some types;

- Direct methane protonic ceramic fuel cell (operating temperature 500 °C).
- Direct methane solid oxide fuel cell (operating temperature 600 °C).
- Direct methane fuel cell with ceria-based anode (operating temperature 650 °C).

5.6 Application of Ammonia in IC Engines

There are two primary phases to the ammonia research activity for internal combustion engines. The early years, from the start of the 1960s to the end of the 1970s, and the later years, starting at the start of the new century. The main difference between the two eras is in the specific goals of the study. Early on, scientists and engineers concentrated on creating engines that could run on alternate fuels in case of another oil crisis. However, the study's primary goal during the second stage is to reduce greenhouse emissions by substituting alternative carbon-free energy sources for some of the fossil fuels.

Ammonia is also a cost-effective alternative to other petroleum-based fuels. With a higher-octane number than natural gas or gasoline, NH_3 makes possible greater compression ratios (CR), improving engine efficiency. Chemically stored ammonia has an energy density significantly higher than that of compressed natural gas and both gaseous and liquid hydrogen, while being comparable to gasoline and diesel

(Valera-Medina et al. 2018). Ammonia is becoming increasingly popular as a low-carbon fuel, and in the next ten years, the market is expected to rise by up to 3%, despite ammonia being classified as a conventional fuel. Spark ignition (SI) engines may successfully run on ammonia, as evidenced by the extensive literature on the subject. A high-octane number is advantageous for ammonia because it can enhance combustion and lessen engine knocking and other undesired combustion effects that are observed in gasoline SI engines. However, because of the fuel's strong resistance to autoignition, burning ammonia in compression ignition engines is a difficult operation.

5.6.1 Combustion, Performance, and Emission Characteristics of IC Engine

The various properties of ammonia's combustion as a fuel give a complicated mixture of profits and disadvantages also. Ammonia's low flame speed is slower than that of conventional hydrocarbon fuels like gasoline or diesel, which is one of its main disadvantages. Slower flame velocity has the tendency to cause incomplete combustion in the engines, which can minimize engine efficiency and increase unburned ammonia emissions, both of which can be very risky (Lif and Holmberg 2006; Tornatore et al. 2022). Ammonia also takes high energy to ignite as traditional fuels because of its high ignition temperature. To properly empower the combustion, this regularly calls for changes like increased and high compression ratios, complicated ignition systems, or the usage of supplemental fuels. Even with these disadvantages, ammonia also has many noteworthy advantages. Rather, there are no direct CO_2 emissions; instead, the main byproducts are water and nitrogen.

The combustion process is crucial to prevent the formation of nitrogen oxides (NO_x), which are pollutants that can contribute to smog and acid rain. Therefore, while ammonia has the potential to be a cleaner alternative fuel, optimizing its use in engines requires addressing its unique combustion characteristics and mitigating its environmental impact.

Ammonia can be used in many applications, which includes ICEs, fuel cells, and as a hydrogen carrier for fuel cell vehicles. Each application has one specific performance characteristic. Suppose direct ammonia cells are efficient and make low emissions but have issues such as catalyst poisoning and ammonia cracking (Shen et al. 2021; Spatolisano et al. 2023). ICEs can use ammonia with some fixtures, but sincere management of combustion parameters is needed to optimize performance and reduce NO_x emissions. Analyzing ammonia as a fuel entails assessing its efficiency and influences on the environment. Because of its high hydrogen concentration, ammonia is seen as a viable fuel for energy storage and transportation. The chemical reaction for ammonia combustion can be summarized as (Eq. 5.6):

$$4NH_3 + 3O_2 \rightarrow 2N_2 + 6H_2O \tag{5.6}$$

This shows that ammonia combusts with oxygen to form nitrogen and water vapor, releasing energy in the process. When comparing ammonia to traditional hydrocarbon fuels, few factors are important. Firstly, ammonia combustion generates zero-carbon emissions, making it an eco-friendlier alternative (Kobayashi et al. 2019; Jafar et al. 2024). However, its energy density is much lower than that of conventional fuels like gasoline or diesel, and it requires more ammonia to produce equal energy and power.

The dual-fuel operation of ammonia combustion is possible, as the pair can be ignited by a secondary fuel that features a much lower autoignition temperature. Diesel was one of the fuels tested in 1966 to become ignited as a dual-fuel engine (Mintz et al. 2014; Hall and Kassa 2021). The 35:1 that the ammonia-only option required was not necessary for a diesel-ammonia dual-fuel compression ignition engine; it could run with as little as 15.2:1 instead. This cetane number of the second fuel is critical in determining the minimum compression ratio needed for a stable operation. In their case, ammonia must be injected within 40 crank angle degrees before diesel injection for good combustion. The combustion of ammonia can promote the formation of nitrogen oxides (NO_x), which have harmful and toxic pollutants (Singh et al. 2022). Effective catalytic converters and advanced combustion technologies are crucial to decrease NO_x emissions.

5.7 Conclusions

It has been observed that ammonia combustion technologies in ICEs, where H_2 is the most common additive with NH_3 in SI engines, enhance combustion efficiency and reduce emissions due to hydrogen's high flame speed and ammonia's high energy density, with the flexibility of using ammonia as a hydrogen carrier or blending it with hydrogen to optimize performance and minimize NOx formation in SI engines. In CI engine, research must be done about the ammonia injection timing. Engine heat rate also improves with aqueous ammonia and emission performances, but it also increases the noise level. Many things still must be found regarding the emission of NO_2 and fuel lean operation. It has been shown that the combustion and emission performances of NH_3 can be improved by the innovation of combustion technologies with observing advancement and cost-effective ammonia, which can be the best production technology based on renewable resources. However, the practical implementation faces hurdles such as ammonia's lower energy density compared to traditional fuels, challenges in ignition, and the potential for NO_x emission. Further research and development are needed to optimize engine design and fuel handling systems to overcome these obstacles and make ammonia a viable and environmentally friendly fuel for ICEs.

Integrating the decarbonization of the power-generating, automobile, and maritime industries to become fossil fuel-independent is possible. During the next several decades, the ICEs may still contribute significantly and cleanly to achieving this objective by operating on carbon-free fuels like ammonia. Liquid ammonia is safer and more readily available for distribution, storage, and transit than hydrogen.

Several nations have the infrastructure necessary to transfer huge amounts safely, and well-established protocols for doing so are recorded.

Acknowledgements The authors would like to express their gratitude to the Royal Academy of Engineering for funding this work through the Transforming Systems through Partnership 23/25 (India) program (TSP-2325–5-IN\145). Their support was instrumental in the successful completion of this study. The fellowship for Dr. Krishna Kumar Pandey is supported by SERB-National Post-Doctoral Fellowship (N-PDF) program and their support is gratefully acknowledged.

References

Adeli K, Nachtane M, Faik A, Saifaoui D, Boulezhar A (2023) How green hydrogen and ammonia are revolutionizing the future of energy production: a comprehensive review of the latest developments and future prospects. Appl Sci 13(15):8711. https://doi.org/10.3390/app13158711

Araia A, Wang Y, Robinson B, Jiang C, Brown S, Wildfire C, Shekhawat D, Hu J (2022) Microwave-assisted ammonia synthesis over Cs-Ru/CeO_2 catalyst at ambient pressure: effects of metal loading and support particle size. Catal Commun 170:106491. https://doi.org/10.1016/j.catcom.2022.106491

Aziz M, TriWijayanta A, Nandiyanto ABD (2020) Ammonia as effective hydrogen storage: a review on production, storage and utilization. Energies (Basel), **13**(12). https://doi.org/10.3390/en13123062

Ben-Iwo J, Manovic V, Longhurst P (2016) Biomass resources and biofuels potential for the production of transportation fuels in Nigeria. Renew Sustain Energy Rev 63:172–192. https://doi.org/10.1016/j.rser.2016.05.050

Chong CT, Hochgreb S (2011) Measurements of laminar flame speeds of liquid fuels: jet-A1, diesel, palm methyl esters and blends using particle imaging velocimetry (PIV), measurements of laminar flame speeds of liquid fuels: Jet-A1, diesel, palm methyl esters and blends using particle imaging velocimetry (PIV), pp 979–986

Dimitriou P, Javaid R (2020) A review of ammonia as a compression ignition engine fuel. Int J Hydrogen Energy 45(11):7098–7118. https://doi.org/10.1016/j.ijhydene.2019.12.209

Edwards TM, Puglis HJ, Kent DB, Durán JL, Bradshaw LM, Farag AM (2024) Ammonia and aquatic ecosystems—a review of global sources, biogeochemical cycling, and effects on fish. Sci Total Environ 907:167911. https://doi.org/10.1016/j.scitotenv.2023.167911

Erdemir D, Dincer I (2021) A perspective on the use of ammonia as a clean fuel: challenges and solutions. Int J Energy Res 45(4):4827–4834. https://doi.org/10.1002/er.6232

Eyidogan M, Ozsezen AN, Canakci M, Turkcan A (2010) Impact of alcohol-gasoline fuel blends on the performance and combustion characteristics of an SI engine. Fuel 89:2713–2720

Faria JA (2021) Renaissance of ammonia synthesis for sustainable production of energy and fertilizers. Curr Opin Green Sustain Chem 29:100466. https://doi.org/10.1016/j.cogsc.2021.100466

Ghavam S, Vahdati M, Wilson IAG, Styring P (2021) Sustainable ammonia production processes. Front Energy Res 9. https://doi.org/10.3389/fenrg.2021.580808

Giddey S, Badwal SPS, Kulkarni A (2013) Review of electrochemical ammonia production technologies and materials. Int J Hydrogen Energy 38(34):14576–14594. https://doi.org/10.1016/j.ijhydene.2013.09.054

Golchoobi A, Tasharrofi S, Taghdisian H (2020) Functionalized nanoporous graphene membrane for water desalination; effect of feed salinity on permeability and salt rejection, a molecular dynamics study. Comput Mater Sci 172:109399. https://doi.org/10.1016/j.commatsci.2019.109399

Hall C, Kassa M (2021) Advances in combustion control for natural gas-diesel dual fuel compression ignition engines in automotive applications: a review. Renew Sustain Energy Rev 148:111291. https://doi.org/10.1016/j.rser.2021.111291

Ishaq H, Crawford C (2024) Review and evaluation of sustainable ammonia production, storage and utilization. Energy Convers Manag 300. https://doi.org/10.1016/j.enconman.2023.117869

Jafar U, Nuhu U, Khan WU, Hossain MM (2024) A review on green ammonia as a potential CO_2 free fuel. Int J Hydrogen Energy 71:857–876. https://doi.org/10.1016/j.ijhydene.2024.05.128

Kobayashi H, Hayakawa A, Somarathne KDKA, Okafor EC (2019) Science and technology of ammonia combustion. Proc Combust Inst 37(1):109–133. https://doi.org/10.1016/j.proci.2018.09.029

Kumar L, Sleiti AK (2024) Systematic review on ammonia as a sustainable fuel for combustion. Renew Sustain Energy Rev 202:114699. https://doi.org/10.1016/j.rser.2024.114699

Kumar S, Agarwal AK, Khandelwal B, Singh P (eds) (2024) Ammonia and hydrogen for green energy transition. Springer, Springer Nature Singapore, Imprint

Liao Z, Zhang S, Zhao Y, Qiu Z, Li K, Han D, Zhang G, Habetler TG (2020) Experimental evaluation of thermolysis-driven gas emissions from LiPF6-Carbonate electrolyte used in lithium-ion batteries. J Energy Chem 49:124–135. https://doi.org/10.1016/j.jechem.2020.01.030

Lif A, Holmberg K (2006) Water-in-diesel emulsions and related systems. Adv Colloid Interface Sci 123–126:231–239. https://doi.org/10.1016/j.cis.2006.05.004

MacFarlane DR, Cherepanov PV, Choi J, Suryanto BHR, Hodgetts RY, Bakker JM, Ferrero Vallana FM, Simonov AN (2020) A roadmap to the ammonia economy. Joule 4(6):1186–1205. https://doi.org/10.1016/j.joule.2020.04.004

Mekhilef S, Saidur R, Safari A (2012) Comparative study of different fuel cell technologies. Renew Sustain Energy Rev 16(1):981–989. https://doi.org/10.1016/j.rser.2011.09.020

Minh NQ (1993) Ceramic fuel cells. J Am Ceram Soc 76(3):563–588. https://doi.org/10.1111/j.1151-2916.1993.tb03645.x

Mintz M, Han J, Burnham A (2014) Alternative and renewable gaseous fuels to improve vehicle environmental performance. In: Alternative fuels and advanced vehicle technologies for improved environmental performance. Elsevier, pp 90–116. https://doi.org/10.1533/9780857097422.1.90

Montoya JH, Tsai C, Vojvodic A, Nørskov JK (2015) The challenge of electrochemical ammonia synthesis: a new perspective on the role of nitrogen scaling relations. Chemsuschem 8(13):2180–2186. https://doi.org/10.1002/cssc.201500322

Reiter AJ, Kong SC (2011) Combustion and emissions characteristics of compression-ignition engine using dual ammonia-diesel fuel. Fuel 90(1):87–97. https://doi.org/10.1016/j.fuel.2010.07.055

Shen H, Choi C, Masa J, Li X, Qiu J, Jung Y, Sun Z (2021) Electrochemical ammonia synthesis: mechanistic understanding and catalyst design. Chem 7(7):1708–1754. https://doi.org/10.1016/j.chempr.2021.01.009

Singh P, Berwal P, Khandelwal B, Kumar S (2022) Impact of ammonia blended fuels on NOx emissions in combustion engine. In: AIAA SCITECH 2022 forum, p 1497

Spatolisano E, Pellegrini LA, de Angelis AR, Cattaneo S, Roccaro E (2023) Ammonia as a carbon-free energy carrier: NH_3 cracking to H_2. Ind Eng Chem Res 62(28):10813–10827. https://doi.org/10.1021/acs.iecr.3c01419

Sun J, Tang Q, Wen M, Huang L, Liu H, Yao M (2024) Combustion characteristics and flame development of ammonia in an optical spark-ignition engine. Fuel 375:132601. https://doi.org/10.1016/j.fuel.2024.132601

Tawalbeh M, Murtaza SZM, Al-Othman A, Alami AH, Singh K, Olabi AG (2022) Ammonia: a versatile candidate for the use in energy storage systems. Renew Energy 194:955–977. https://doi.org/10.1016/j.renene.2022.06.015

Tornatore C, Marchitto L, Sabia P, De Joannon M (2022) Ammonia as green fuel in internal combustion engines: state-of-the-art and future perspectives. Front Mech Eng 8. https://doi.org/10.3389/fmech.2022.944201

Valera-Medina A, Xiao H, Owen-Jones M, David WIF, Bowen PJ (2018) Ammonia for power. Prog Energy Combust Sci 69:63–102. https://doi.org/10.1016/j.pecs.2018.07.001

Verschoor JC, de Jongh PE, Ngene P (2024) Recent advances in thermocatalytic ammonia synthesis and decomposition. Curr Opin Green Sustain Chem 100965

Wang L, Ji S, Wang N, Zhang R, Zhang G, Li JR (2014) One-step self-assembly fabrication of amphiphilic hyperbranched polymer composite membrane from aqueous emulsion for dye desalination. J Membr Sci 15(452):143–151

Withagen C (1994) Pollution and exhaustibility of fossil fuels. Resour Energy Econ 16(3):235–242. https://doi.org/10.1016/0928-7655(94)90007-8

Yang D, Li J, Xiao M, Liu C, Xing W, Zhu J (2023) Atomically dispersed metal catalysts towards nitrogen reduction for ammonia: from homogeneous to heterogeneous. Chem Eng J 468:143776. https://doi.org/10.1016/j.cej.2023.143776

Yapicioglu A, Dincer I (2019) A review on clean ammonia as a potential fuel for power generators. Renew Sustain Energy Rev 103:96–108. https://doi.org/10.1016/j.rser.2018.12.023

Zamfirescu C, Dincer I (2008) Using ammonia as a sustainable fuel. J Power Sources 185(1):459–465. https://doi.org/10.1016/j.jpowsour.2008.02.097

Zhang J, Lin L, Cui H (2020) Plasma-assisted synthesis of platinum nitride nanoparticles under HPHT: realized by carbon-encapsulated ultrafine Pt nanoparticles. Nanomaterials 10(9):1780. https://doi.org/10.3390/nano10091780

Zhang K, Shang S, Li X, Gao W (2023) Lower flammability limits of NH_3/H_2 mixtures under different initial temperatures and initial pressures. Fuel 331:125982. https://doi.org/10.1016/j.fuel.2022.125982

Zhu N, Zhang K, Wu F, Bai Y, Wu C (2021) Ionic liquid-based electrolytes for aluminum/magnesium/sodium-ion batteries. Energy Mater Adv

Chapter 6
Hydroxy Gas as a Substitute to Diesel Fuel for Sustainable Development

Nidhi Sharma

Abstract Exploring a diesel substitute for compression ignition engine is a crucial step toward achieving cleaner, healthier, and more sustainable transportation and energy systems. Hydroxy (HHO) gas is considered as a partial substitute to diesel fuel because of its combustion characteristics. Hydrogen-based fuel can be generated on-demand through water electrolysis for immediate use, eliminating the need for storage. As a potential substitute for or supplement to diesel fuel, HHO offers a promising solution for sustainable energy development. The depletion of fossil fuels and emissions from automobiles drive climate change, prompting a shift toward cleaner alternatives. Hydrogen (H_2), particularly as HHO-a mixture of hydrogen and oxygen, emerges as a renewable fuel that enhances combustion efficiency in engines. This chapter explores the properties and benefits of hydrogen and HHO, outlines various electrolyzer types, and highlights their role in sustainable energy solutions. Key characteristics of hydrogen are crucial for optimizing engine performance and preventing hazards. The chapter discusses flammability limits, combustion risks, and material challenges like hydrogen embrittlement. It also examines HHO gas production optimization, comparing wet and dry cell electrolyzers. Integrating HHO into diesel engines shows potential in reducing fuel consumption and improving thermal efficiency, offering a pathway for enhanced fuel efficiency and lower environmental impact.

Keywords HHO gas · Hydroxy fuel · Brown's gas · Diesel substitution

6.1 Introduction

The world is competing with both a fossil fuel shortage and the urgent need to mitigate environmental degradation. As global energy demands continue to rise, the depletion of fossil fuel resources is becoming increasingly apparent, posing challenges for

N. Sharma (✉)
Amity University Uttar Pradesh, Noida, Uttar Pradesh, India
e-mail: nsharma18@amity.edu

N. Sharma et al. (eds.), *Sustainable, Scalable and Storable E-Fuels for Decarbonising Transport Sector*, Energy, Environment, and Sustainability,
https://doi.org/10.1007/978-981-96-7388-9_6

energy security and economic stability. Automobiles are the major contributors to these challenges and they emit pollutants such as carbon monoxide (CO), nitrous oxide (NOx), and hydrocarbons (HC). These emissions not only degrades the air quality but also play a pivotal role in the acceleration of global warming, exacerbating climate change and its associated impacts. In response to these challenges, battery-operated vehicles are emerging as the driving force in the automobile industry. These electric vehicles (EVs) offer a cleaner alternative to traditional internal combustion engines (ICEs), which are responsible for much of the pollution from road transport. However, the transition to EVs is complex and requires substantial advancements in technology, infrastructure, and consumer acceptance. While progress is being made, it remains evident that there is still a significant journey ahead before battery-operated vehicles can replace ICEs to some extent. Diesel engines, in particular, continue to play a crucial role in various sectors, including power generation, heavy-duty trucking, and construction equipment. These engines are preferred for their relatively better efficiency and durability, especially in demanding applications. Nevertheless, as awareness of environmental issues grows, the demand to reduce emissions from these engines is steadily increasing. Furthermore, governments and regulatory bodies are increasingly implementing stringent emission standards to combat pollution. These regulations are designed to limit the amount of harmful pollutants that can be released into the atmosphere, compelling manufacturers to innovate and adopt cleaner technologies. As a result, the automobile industry is at a pivotal point, facing both the necessity of adaptation and the opportunity for innovation.

Hydrogen (H_2) is increasingly being recognized as a promising fuel for the future transport sector. It serves as an energy carrier, much like electricity, offering versatility in various applications. Government around the world is initiating programs to promote hydrogen as an alternative energy source, aiming to create sustainable energy systems that can help mitigate climate change. Historically, hydrogen has been produced from fossil fuels, but there is a growing emphasis on generating hydrogen from renewable energy. This shift not only reduces dependence on fossil fuels but also enhances the sustainability of hydrogen production. Hydrogen has numerous applications, including its use in ICE, gas turbines, and fuel cells. Each of these technologies can benefit from hydrogen's clean-burning properties, which produce only water vapor as a byproduct when combusted. Despite its potential, the initial production cost of hydrogen remains high, which poses a barrier to widespread adoption. However, ongoing research and technological advancements are expected to lower these costs, positioning hydrogen as a viable alternative to electricity as an energy carrier in the future. Figure 6.1 illustrates the various methods of hydrogen production, highlighting its capability to be derived from almost all energy sources. This flexibility addresses one of the key challenges of modern energy systems: the limitations of battery technology in automobiles and other applications.

HHO, a stoichiometric mixture of hydrogen and oxygen in a 2:1 ratio, exists in a gaseous state at atmospheric temperature and pressure. Due to its high specific volume, HHO is challenging to handle and cannot be stored easily, necessitating its direct use in engines. Many heavy-duty diesel engines are now equipped with HHO kits to enhance efficiency and reduce fuel consumption. While HHO acts as

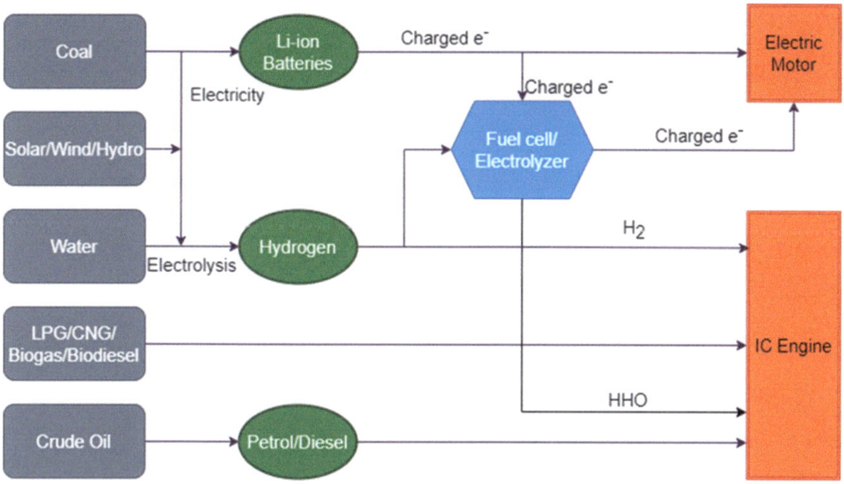

Fig. 6.1 Various sources of energy

a secondary fuel, it must be used alongside diesel or gasoline, which serve as the primary fuels.

In the engine, HHO improves the combustion of the primary fuel by raising the temperature in the combustion chamber and enabling a faster burn due to its high energy density. Traditionally, HHO has been integrated into heavy-duty diesel engines; however, smaller petrol engines are increasingly being fitted with HHO kits to boost fuel economy.

HHO is produced through the electrolysis of water (H_2O). This process separates water into hydrogen (H_2) and oxygen (O_2) using a high-voltage current. This process occurs in an electrolyzer, equipped with electrodes and an electrolyte solution. When a high-voltage current is applied to the electrodes, water is split into hydrogen at the cathode and oxygen at the anode.

In these electrolyzers, the anode and cathode are typically not separated by any medium and are located within the same container. This configuration allows for the generation of a gas mixture commonly referred to as HHO gas. Equation for electrolysis of water is.

- At cathode: $2\ H_2O + 2e^- \rightarrow H_2 + 2OH$
- At anode: $4OH \rightarrow O_2 + 2\ H_2O + 4e^-$
- Overall reaction: $2\ H_2O \rightarrow 2\ H_2 + O_2$

Properties of Hydrogen

Hydrogen possesses several unique physical properties that make it a noteworthy fuel alternative. As the smallest element in the universe, hydrogen has the highest energy density of all gases, being eight times lighter than methane (natural gas). Specifically, one kilogram of hydrogen can store approximately 120 MJ of energy. Table 6.1 shows properties of HHO w.r.t. hydrogen.

Table 6.1 Properties of HHO w.r.t. hydrogen (Paparao and Murugan 2021)

Property	Oxy hydrogen (HHO)	Hydrogen (H_2)
Density (kg/m^3) at 16 °C and 1.01 bar	0.5	0.1
Molecular weight	12	2
Lower heating value (MJ/kg)	13.2	119.9
Higher heating value (MJ/kg)	25.9	142.2
Flame velocity (cm/s)	265–325	265–325
Diffusivity in air (cm^2/s)	0.6	0.6
Flammability limits (vol % in air)	4–95	4–75

Key Physical Properties of Hydrogen:

- Appearance: Colorless, odorless, and tasteless gas; non-poisonous.
- Molar Mass: 2 g/mol.
- Calorific Value: 142 MJ/kg.
- Boiling Point: −252 °C.

Hydrogen is particularly suitable as an alternative fuel for internal combustion (IC) engines due to several advantageous properties:

- Wide flammability limits.
- High calorific value.
- High laminar flame speed.
- Short quenching distance.
- High diffusivity.

Moreover, hydrogen combustion produces no smoke, sulfur compounds, lead, or phosphorus compounds, resulting in virtually colorless emissions.

Advantages of HHO Gas:

- Sustainable Energy Source: HHO can be generated from water, making it widely accessible. It reduces harmful exhaust emissions, such as hydrocarbons (HC), CO, and particulates, due to its "clean" burning characteristics.
- Lightweight: HHO gas is very light.
- Wider Flammability Limit: This allows for more efficient combustion.
- Rapid Dispersion: HHO gas spreads quickly in the combustion chamber, promoting even fuel combustion.
- Improved Combustion Efficiency: Its high energy density enhances the combustion of the primary fuel when mixed.
- Zero CO_2 Emissions: Combustion of HHO gas produces only water vapor as a byproduct, eliminating carbon emissions.
- HHO can be easily produced in compact setups, which mitigates the challenges associated with storage and transportation.

Drawbacks of HHO Gas:

- Energy Carrier: HHO is an energy carrier and must be produced from an external source.
- Electrolysis Requirements: The electrolysis of water to produce HHO requires a high-voltage current.
- Low Boiling Point: Its boiling point is −252 °C. The low boiling point gives significant storage challenges.
- Storage Difficulties: Storing HHO gas requires cryogenic cooling or high-pressure conditions, complicating logistics.
- NOx Formation: The reaction of O_2 in HHO with N_2 in the atmosphere can lead to the formation of nitrogen oxides (NOx).
- Thermal Efficiency Issues: It may reduce thermal efficiency due to the higher energy content input into the combustion chamber.

Basic Types of HHO Gas Generators:

HHO generators are primarily categorized into two types: (i) wet cell and (ii) dry cell, with dry cells further divided into three types: (a) alpha, (b) beta, and (c) omega (Subramanian and Ismail 2018; Ismail et al. 2018). Wet cell generators feature electrodes submerged in a water and electrolyte solution, while dry cell generators enhance the design by modifying the arrangement of the electrolyte reservoir and electrodes (Paparao and Murugan 2021).The key components of dry cell HHO generators include electrodes, an electrolyte, a reservoir, a bubbler tank, fittings, non-conductive side plates, and assembly hardware, as illustrated in Fig. 6.2. Common electrode materials are iron, stainless steel, graphite, copper, and aluminum. Typical electrolytes include acetic acid, sodium bicarbonate, sodium hydroxide, potassium hydroxide, and sodium chloride. The dry cell HHO gas reactor is shown in Fig. 6.3. The benefits of dry cell generators include lower current requirements, a more compact design, less maintenance, and reduced corrosion. Figure 6.4 shows wet cell HHO generators.

Overall, HHO gas is crucial for promoting sustainable energy solutions and combating climate change. As a clean energy source, hydrogen helps reduce reliance on fossil fuels and minimizes greenhouse gas emissions. Understanding its applications can drive innovation in technology, enhance energy security, and create economic opportunities.

6.2 Hydrogen/HHO Physical and Chemical Characteristics

Understanding the physical and chemical properties of H_2/HHO is essential, as they significantly influence engine performance. The leakage of hydrogen and oxygen poses critical challenges for the operation of high-pressure water electrolyzers and fuel cells. Such contamination can lead to malfunctions, and in severe cases, it can result in explosions within gas cylinders. Therefore, it is important to understand the properties of hydrogen-air and hydrogen–oxygen mixtures to prevent hazardous situations in a engine test cell.

Fig. 6.2 Various types of HHO gas generator

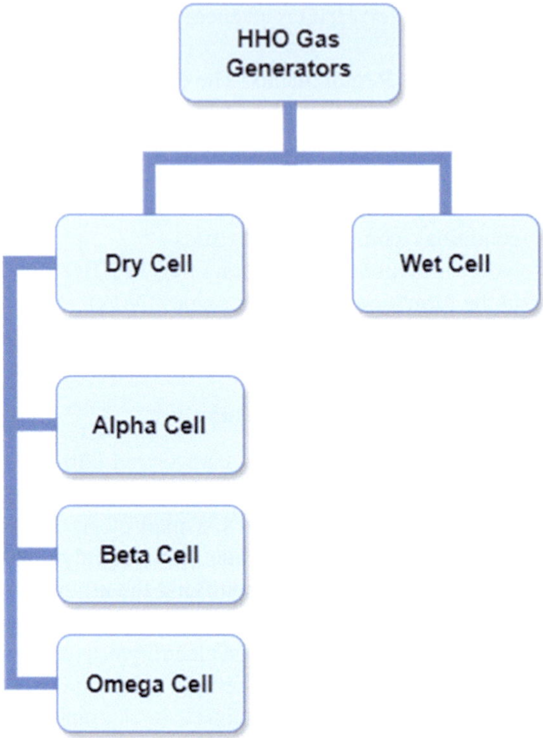

Fig. 6.3 Dry cell HHO gas reactor

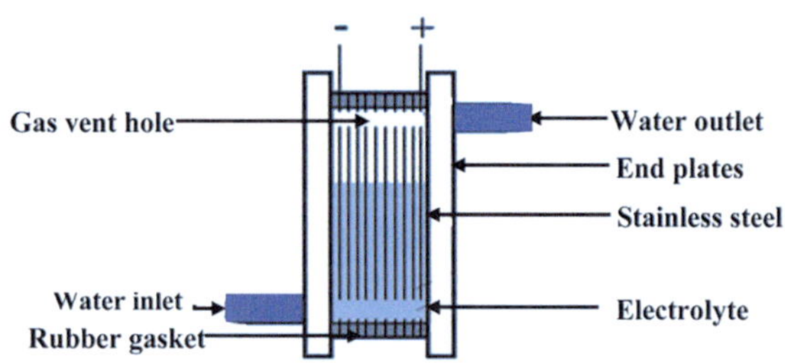

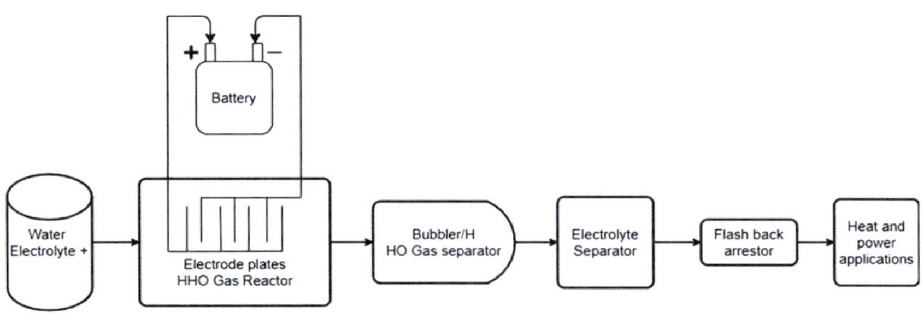

Fig. 6.4 Wet cell HHO gas generator

Physical Properties

This section examines how various parameters affect knocking and backfiring (or pre-ignition) in a single cylinder hydrogen-fueled engine. Different H_2-combustion engines have been developed based on the physical and chemical characteristics of hydrogen/HHO gas and its combination with air.

Flammability: The flammability limits of hydrogen range from lean mixtures ($\lambda = 10$) to rich mixtures ($\lambda = 0.14$), expanding as temperature rises. The lower flammability limit increases with higher pressure.

Uncontrolled Combustion: While hydrogen's wide flammability limits, low ignition energy requirements, and fast flame speeds make it suitable for ICE these same characteristics can also lead to abnormal combustion events, such as surface ignition, backfiring, and autoignition.

Surface Ignition: This uncontrolled ignition occurs due to hot spots in the combustion chamber. To mitigate this risk, it is important to avoid or reduce these hot areas, as they can lead to surface ignition or backfiring. Hydrogen engines can operate without a throttle valve, which can reduce pumping losses. Proper engine design can significantly decrease the likelihood of surface ignition.

Backfire: Backfiring happens when a fresh charge ignites in the combustion chamber during the intake stroke. As the intake valves open, a new hydrogen-air mixture is drawn in, but autoignition of remaining end gas can occur later in the combustion process. Delayed spark timing increases backfiring risk due to higher exhaust temperatures and residual in-cylinder temperatures. Backfiring can be minimized by using direct injection and a well-timed manifold injection system to prevent unwanted combustion events.

Pre-ignition: Pre-ignition refers to combustion that starts before the intended spark timing, occurring during the compression stroke. It becomes more pronounced when hydrogen-air mixtures reach stoichiometric levels, particularly under high engine speed and load conditions. Direct injection helps prevent pre-ignition in the intake manifold, although it may still occur in the combustion chamber. A well-designed direct injection system tailored to specific engine configurations is essential for smooth operation.

Autoignition: Autoignition occurs when end-gas conditions (pressure and temperature) cause the remaining gas to ignite spontaneously. This rapid energy release creates high-amplitude pressure waves, known as engine knock. Severe knocking increases mechanical and thermal stress, potentially damaging the engine.

Knocking: Knocking in ICE results from abnormal combustion. In compression ignition (CI) engines, knocking can occur at the start of combustion, where a short delay between fuel injection and ignition leads to rapid burning and a sudden pressure increase, producing a metallic hammer-like sound known as knocking.

Chemical Properties

Hadi et al. (2019) demonstrated the effects of hydrogen embrittlement on the mechanical properties of iron and steel. The primary consequences of hydrogen embrittlement are a loss of ductility and reduced fracture stress. This embrittlement occurs due to a chemical reaction that produces hydrogen on the material's surface. In hydrogen-rich environments, environmental brittleness arises from the adsorption of molecular hydrogen, which can then transform into atomic form and accumulate within the material's lattice. This phenomenon limits material selection for hydrogen fuel and storage systems due to the ecological embrittlement in hydrogen-containing atmospheres (Hadi et al. 2019).

6.3 Optimization of HHO Gas with IC-Engine

This topic investigates whether the HHO electrolyzer can meet the requirements of a diesel engine. Researchers examined the effects of HHO produced by both wet and dry cell electrolyzers, as well as the feasibility of using the HHO gas system for on-site applications.

Construction of Cell

Researchers worldwide have constructed the electrolyzer cell and studied the effects of various design characteristics using the design of experiments (DOE) methodology. The cell was developed by Manu et al. (2016), utilizing 316L stainless steel for the electrodes and neoprene for the gasket. Key parameters included a current density of 0.1 A/cm^2, electrode spacing of 2 and 3 mm, and the use of NaOH and KOH as electrolyte types. Additionally, the electrolyte concentrations tested were 20 and 30%, with variations in electrode surface texture (smooth versus rough) (Manu et al. 2016). Subramanian and Thangavel (2020a) experimented with varying concentrations of NaOH in the solution, testing levels of 0.25, 0.5, 0.75, and 1 Molarity. They found that conductivity increased significantly with higher electrolyte concentrations, while the electrical potential requirement decreased. Subramanian and Thangavel (2020a). Manu et al. (2016) determined the minimum voltage required to initiate electrolysis between adjacent plates, identifying KOH as the most effective electrolyte. They concluded that a spacing of 3 mm should be maintained between the electrodes, with an optimal electrolyte concentration of 30% and rough electrode surfaces yielding the best results. Under these conditions, the minimum voltage required per gap was calculated to be 1.829 V (Manu et al. 2016).

Rousan et al. (2010) found that injecting HHO gas into the intake manifold reduced fuel consumption in a gasoline engine. In their experiment, HHO gas was introduced into the air stream just before entering the carburetor. They determined that the optimal size for the fuel cell should have an electrolyte surface area 20 times larger than the piston's surface area to generate sufficient HHO. Additionally, the amount of water required in the cell was approximately half of the engine's capacity (Al-Rousan 2010). According to Gohar and Hassan (2017), the electrolysis of water in the HHO generator yields approximately 55% hydrogen and 45% oxygen, whereas the ideal proportions are 66.6% hydrogen and 33.3% oxygen. They identified KOH as the most effective catalyst, which produced 5 L of gas per minute (Gohar and Hassan 2017). Kady et al. (2020) produced HHO gas using an electrolyzer with stainless steel 316L plate electrodes and alkaline water. They investigated the effects of electrolyte concentration, cell design, electric current, voltage, time, and operating temperature. The dry cells consisted of 26 stainless steel 316L plates, each with a surface area of 136.5 cm^2, and a spacing of 4 mm between each pair of plates. Electrolyte concentrations varied from 5 to 20% for both KOH and NaOH.

The study revealed several findings: higher applied voltage and current led to increased gas production. Using tap water resulted in lower current usage and reduced HHO output. Neutral plates improved the effective area while decreasing the current supplied. For the same configuration and electrolyte content, NaOH was preferred over KOH due to its lower molecular weight.

The specified current was increased from 6 to 42 A, which raised the gas flow rate to 654 ml/min. The electrolyzer's efficiency increased by 72.9% when the current rose from 6 to 14 A, although efficiency declined after 30 min of operation. With electrolyte concentrations of 5, 10, 15, and 20% NaOH and flow currents of 14, 18, 20, and 21.3 A, gas flow rates of 866, 985, 1040, and 1090 ml/min were achieved. Additionally, the electrolyzer efficiency and HHO flow rate increased to 51.12% and 246 ml/min, respectively, as the applied voltage was raised from 2 to 4 V (Kady et al. 2020).

Comparison of Wet and Dry Cell

Numerous types of electrolyzers are available on the market, with wet and dry cell electrolyzers being the most efficient. Soly et al. (2021) conducted comparative experimental research on the HHO generation rates of both types. They investigated the impacts of several factors on HHO production, including electrolyte content, cell configuration, electric current, operating time, electrolyte temperature, and voltage.

The findings revealed that wet cells produce more HHO gas than dry cells under the same design specifications, as all electrodes are submerged in the electrolyte solution, increasing the active surface area. However, wet cells generate more heat and occupy more space than dry cells. Dry cells are preferred for their smaller size, simpler manufacturing process, and ease of assembly. While wet cells are generally less expensive, they produce more HHO gas than dry cells. Additionally, dry cells tend to generate less heat and experience less electrode corrosion compared to wet cells (Soly et al. 2021). Gad and Razek (2021) compared dry and wet cells, finding that the maximum reductions in brake specific fuel consumption (BSFC) were 2.5 and 3%, respectively. At all speeds and full load, the highest output power gains for dry and wet cells were 3 and 4%, respectively, compared to diesel. The most significant increases in thermal efficiency were 2 and 2.5% for dry and wet cells at all speeds and full load. Consequently, dry cells were preferred over wet cells due to their lower cost, smaller size, and easier manufacturing and assembly. In contrast, wet cells produce more heat and waste energy, occupy more space, and experience higher electrode corrosion than dry cells. Kady et al. (2020) focused on modifying the design to maximize HHO gas generation. They created HHO using a low-cost electrolyzer made from readily available components. The dry cell type was preferred over the wet cell type because the electrolyte fills the electrode gap and reduces the generated heat through the circulation of hot and cold water (Kady et al. 2020).

6.4 Effect of HHO Gas on Engine Performance and Combustion

This section examines the characteristics of a hydroxy-diesel CI system. The hydrogen dual-fuel engine can operate on lean fuel mixtures during part load and no load conditions. By using blended fuels, the engine derives most of its energy from diesel, with the remainder coming from hydrogen. According to Yilmaz et al. (2010) the HHO system combined with diesel fuel produces better engine torque at mid to high engine speeds compared to conventional diesel operation. The rapid burning velocity and low ignition energy of the hydroxy-air mixture reduce in-cylinder charge flow and increase the percentage of residual gas, which can hinder the complete and rapid combustion of fuel at high speeds (Yilmaz et al. 2010). Yadav Milind et al. (2011) reported that blending oxyhydrogen gas with diesel led to a reduction in BSFC. During testing, the engine required nearly 10 g less fuel to produce the same amount of power per hour at a maximum load of 8 kg. This reduction could be further improved by increasing the delivery of oxyhydrogen gas. The substitution of oxyhydrogen for diesel during combustion results in a decrease in BSFC, as oxyhydrogen contributes some of the necessary input energy, thereby reducing the amount of fuel needed to generate equivalent energy. When a fuel mixture is injected into the combustion chamber, a greater volume of the mixture reaches the chamber than the engine's swept volume, enhancing volumetric efficiency (Yadav Milind et al. 2011). Saravanan et al. (2008a) utilized exhaust gas recirculation (EGR) technology to improve the engine's volumetric efficiency, resulting in increased brake power. The inert gas from the exhaust gas recirculation (EGR) system helps lower peak combustion temperatures, which in turn reduces NOx emissions. Additionally, this approach enhances fuel economy and decreases pollutant emissions. Overall, the implementation of EGR technology led to a significant reduction in NOx emissions (Saravanan et al. 2008a).

The performance of a hydrogen-fueled direct injection engine was investigated, focusing on hydrogen-enriched air as the intake charge in a CI engine operating in dual-fuel (D.F.) mode. Gohar and Hassan (2017) evaluated the effects of varying HHO levels on brake thermal efficiency (BTE) and brake specific energy consumption (BSEC) under different engine load conditions. Their findings revealed that using HHO gas resulted in an average reduction of 27% in SFC. This improvement was due to the uniform mixing of HHO with air-thanks to its high diffusivity-and the enhanced oxygen content of HHO gas, that relatively improved diesel combustion and performance characteristics (Gohar and Hassan 2017). Subramanian and Thangavel (2020b) delivered HHO gas at flow rates of 6, 12, 18, 24, 30, and 36 L per minute (LPM) into the intake manifold along with air. They found that at zero torque, a flow rate of 36 LPM of HHO gas achieved the highest energy substitution, reaching 76% (Subramanian and Thangavel 2020b). Yadav et al. (2014) found that when operated on hydrogen, the maximum efficiency reached approximately 70% at full load, whereas when using diesel, the maximum efficiency was nearly 80% at full load. Notably, BTE increased by 11.6 at 70% full load. Sharma et al. (2020)

delivered HHO gas with diesel at three different flow rates: 0.25, 0.50, and 0.75 lpm. At 0.75 lpm, the HHO flow rate and 10 kg load increased 6.5 and 6%, respectively. Under DF mode, under the same load and flow rate conditions (Sharma et al. 2020). Bose and Maji (2009) investigated the effect of hydrogen enrichment (0.15 kg/h hydrogen flow rate) combined with 10 and 20% Exhaust Gas Recirculation (EGR) on engine performance. They conducted tests under various load conditions: no load, 20% load, 40% load, 60% load, and 80% load. The study concluded that without EGR, BTE increased by 12.9% with a hydrogen flow rate of 0.15 kg/h. (Bose and Maji 2009). Bari and Esmaeil (2010) reported that BTE improved by 2.6% at 19 kW, 2.9% at 22 kW, and 1.6% at 28 kW when a 6.1% diesel-equivalent H_2/O_2 mixture was injected into the diesel engine. Additionally, BSFC decreased by 7.3, 8.1, and 4.8% at 19, 22, and 28 kW, respectively. However, HHO additions exceeding 5% had no noticeable effect on engine performance (Bari and Esmaeil 2010). Wang et al. (2012) used a heavy-duty diesel engine to study the effects of varying quantities of HHO mixture. They found that BTE increased from 31.1% with diesel alone to 39.9% with the addition of HHO. The BSFC was higher than that of diesel when using 10–40 LPM of HHO. However, at HHO addition rates of 50, 60, and 70 LPM, BSFC was approximately 3.2, 9.9, and 10.5% lower than diesel, respectively (Wang et al. 2012).

Several studies have investigated the performance effects of HHO on different fuels. Chiriac et al. (2006) found that in gasoline (SI) engines, HHO mixed with controlled pressure and temperature increased BTE by up to 1.2% and Net IMEP by 5.6% at a lambda (λ) ratio of 0.92–0.94. At $\lambda = 1.18–1.2$, these values improved further, with efficiency and IMEP rising by up to 3 and 17%, respectively. Yilmaz et al. (2010) reported a 19% increase in engine torque and a 14% reduction in SFC in diesel (CI) engines supplied with HHO. Al-Rousan (2010) introduced HHO into gasoline (SI) engines before the carburettor, achieving a 6% increase in thermal efficiency and a 20% reduction in fuel consumption. Bari and Esmaeil (2010) noted a 3% increase in thermal efficiency in diesel (CI) engines, along with a 10% reduction in total fuel consumption and a 15% decrease in diesel use.

The study aims to explore the combustion characteristics of hydrogen and compare them to those of diesel. The effects of adding a small amount of HHO mixture as an additive to a four-cylinder diesel engine were investigated by Yadav et al. (2014) based on the experimental findings, the following conclusions were drawn: the optimal injection timing for a hydrogen-fueled engine was found to be 20° CA before top dead center (BTDC) with a hydrogen flow rate of 120 g/h. (Yadav et al. 2014). Bose and Maji (2009) investigated the use of hydrogen enrichment (0.15 kg/h hydrogen flow) combined with 10 and 20% EGR to operate the engine. They found that at high rates of hydrogen addition, combustion became uncontrollable, leading to a decrease in thermal efficiency. Without EGR, the brake specific energy consumption (BSEC) was lower than that of conventional diesel operation. This reduction in BSEC is attributed to hydrogen's higher calorific value and the lean-burn conditions under which hydrogen-fueled engines operate (Bose and Maji 2009). Saravanan et al. (2008b) conducted experiments with hydrogen in dual-fuel mode on a diesel engine. The optimal conditions in dual-fuel mode were found to be a hydrogen injection timing of 5° crank angle (CA) BTDC and an injection duration of 30° CA. However, due to the rapid combustion of hydrogen at high loads, significant knocking was observed when the engine operated above 75% load with hydrogen (Saravanan et al. 2008b). Bari and Esmaeil (2010) considered the potential of on-board HHO mixture generation, which has a flame speed nine times faster than diesel. This higher flame speed could enhance overall combustion by generating higher peak pressure closer to top dead center (TDC), resulting in increased work output (Bari and Esmaeil 2010). Table 6.2 compares characteristics of hydroxyl gas and diesel fuel.

Summary

- The shift to sustainable energy is essential amid fossil fuel shortages and environmental concerns. Hydrogen (H_2) and HHO offer clean combustion and high energy density, but challenges like production costs and NOx emissions limit adoption. Continued advancements in hydrogen technologies can help reduce greenhouse gas emissions and improve energy security.
- Understanding hydrogen's properties is crucial for optimizing engine performance and safety. Its flammability enhances combustion efficiency but also poses risks that require careful design and material selection.
- HHO electrolyzers can improve diesel engine efficiency, with dry cells preferred for their compactness despite wet cells producing more HHO. Optimizing design parameters is key for practical use.
- Integrating HHO in hydroxy-diesel systems leads to better efficiency and emissions reduction. Hydrogen improves torque and lowers brake specific fuel consumption, while exhaust gas recirculation (EGR) optimizes combustion and reduces emissions. Overall, HHO shows great promise for sustainable diesel engine operations.
- HHO has potential for fuel savings and emission reduction, but more peer-reviewed studies needed for validation. Effectiveness and safety are still debated.

HHO gas shows potential as a viable alternative or supplement to diesel fuel, contributing to cleaner and more efficient transportation solutions. Continued research and development are essential to optimize its use and address any technical challenges.

Table 6.2 Characteristics of hydroxyl gas and diesel fuel

Parameter	Hydroxy gas	Diesel fuel
Composition	Hydroxy gas is a mixture of hydrogen (H_2) and oxygen (O_2), created through the electrolysis of water (H_2O). Example: Electrolysis of water to produce hydrogen and oxygen gases	Diesel fuel is composed mainly of hydrocarbons, primarily alkanes, derived from crude oil
Source	Hydroxy gas is produced by splitting water molecules using electricity, which can be sourced from renewable or non-renewable energy sources	Diesel is sourced from crude oil extracted through drilling, which is a finite and non-renewable resource
	Example: Using solar energy to electrolyze water and produce hydroxy gas	Example: Crude oil extraction from oil rigs in the Middle East
Energy source for production	The production of hydroxy gas requires electricity for the electrolysis process, and if renewable energy (e.g., solar, wind, or hydroelectric) is used, the production process can be considered sustainable	Diesel is refined from crude oil, an energy-intensive process that relies on fossil fuels, contributing to its carbon footprint
	Example: A solar-powered water electrolyzer producing hydroxy gas for fuel	Example: Diesel fuel refined from crude oil in a refinery
Emissions	When hydroxy gas burns, it produces only water vapor and heat, resulting in minimal or zero emissions, making it a clean fuel	Diesel combustion releases a range of pollutants including carbon dioxide (CO_2), nitrogen oxides (NOx), particulate matter, and sulfur compounds, contributing to air pollution and climate change
	Example: A hydrogen-powered fuel cell vehicle emitting only water vapor	Example: Diesel-powered trucks emitting black smoke and pollutants
Environmental impact	Hydroxy gas has a significantly lower environmental impact, especially when produced using renewable energy, as it produces little to no pollutants	Diesel fuel has a high environmental impact, contributing significantly to global warming due to CO_2 emissions and air pollution from other harmful exhaust gases
	Example: A zero-emission vehicle running on hydrogen or hydroxy gas	Example: Diesel-powered cars and trucks emitting large amounts of CO_2

(continued)

Table 6.2 (continued)

Parameter	Hydroxy gas	Diesel fuel
Fuel efficiency	Hydroxy gas can potentially offer higher fuel efficiency in certain engine configurations, though this depends on the technology and optimization of combustion processes	Diesel fuel has high energy density and established fuel efficiency, but some diesel engines may be less efficient in comparison to hydrogen or hydroxy gas-powered engines under certain conditions
	Example: Hydrogen engines used in Formula 1, which are being tested for efficiency	Example: Diesel engines in long-haul trucks offering high fuel efficiency for distance travel
Storage and distribution	Hydroxy gas presents challenges in storage and distribution due to its low energy density and the flammable nature of hydrogen, requiring specialized tanks and infrastructure	Diesel fuel benefits from a well-established, efficient global distribution network, with a high energy density that makes storage and transportation easier and more cost-effective
	Example: Hydrogen storage tanks used in hydrogen-powered cars	Example: Fuel stations worldwide offering diesel fuel for trucks and cars
Engine compatibility	Diesel engines need modifications or new designs to run on hydroxy gas, as standard diesel engines are not compatible with hydrogen or hydroxy gas combustion	Diesel fuel is compatible with the vast majority of internal combustion engines currently in use worldwide, with no significant modifications required
	Example: Modified internal combustion engine (ICE) running on hydrogen in a prototype car	Example: Diesel engines in cars, buses, and trucks that are ready for use without modifications
Cost of production	The production cost of hydroxy gas is currently high, particularly when renewable electricity is used for electrolysis, which can make it less economically viable compared to fossil fuels	Diesel fuel production costs are relatively lower due to established extraction and refining infrastructure, but fluctuating crude oil prices can affect cost stability
	Example: High cost of producing hydrogen via electrolysis using solar power	Example: Diesel refined from crude oil at refineries such as ExxonMobil
Infrastructure needs	Significant new infrastructure is required for the production, storage, and distribution of hydroxy gas, particularly for hydrogen storage and specialized fueling stations	Diesel has a mature infrastructure for extraction, refining, and global distribution, making it widely available and easy to access
	Example: Hydrogen fueling stations such as those being developed in California	Example: Global network of diesel refueling stations for cars, trucks, and ships

(continued)

Table 6.2 (continued)

Parameter	Hydroxy gas	Diesel fuel
Sustainability	Hydroxy gas can be a sustainable fuel, particularly when produced using clean, renewable energy sources, as it is derived from water, a renewable resource	Diesel is not sustainable in the long term, as it is derived from finite fossil fuel reserves and contributes to environmental degradation through emissions and resource depletion
	Example: A solar-powered hydrogen production facility producing sustainable fuel	Example: Ongoing oil extraction from finite reserves in the North Sea
Current commercial use	Hydroxy gas is not yet widely commercially used, with its applications mostly limited to experimental and niche uses, and it is not yet a mainstream fuel option	Diesel fuel is widely used around the world, powering a variety of vehicles, industrial machinery, and other applications, making it a primary source of energy globally
	Example: Prototype hydrogen-powered cars like the Toyota Mirai	Example: Diesel-powered trucks, buses, and heavy machinery in daily use worldwide

List of Abbreviation

BTDC	Before top dead center
BTE	Brake thermal efficiency
CO	Carbon monoxide
ICE	Combustion engines
CI	Compression ignition
CA	Crank angle
DOE	Design of experiments
D.F.	Dual fuel
EVs	Electric vehicles
H_2O	Electrolysis of water
EGR	Exhaust gas recirculation
SI	Spark Ignition
HC	Hydrocarbons
H_2	Hydrogen
HHO	Hydroxy
IC	Internal combustion
LPM	Liters per minute
NOx	Nitrogen oxides
O_2	Oxygen
BSFC	Brake specific fuel consumption
SFC	Specific fuel consumption
TDC	Top dead center

References

Al-Rousan AA (2010) Reduction of fuel consumption in gasoline engines by introducing HHO gas into intake manifold. Int J Hydrogen Energy 35(23):12930–12935
Bari S, Esmaeil MM (2010) Effect of H_2/O_2 addition in increasing the thermal efficiency of a diesel engine. Fuel 89(2):378–383
Bose PK, Maji D (2009) An experimental investigation on engine performance and emissions of a single cylinder diesel engine using hydrogen as inducted fuel and diesel as injected fuel with exhaust gas recirculation. Int J Hydrogen Energy 34(11):4847–4854
Chiriac R, Apostolescu N, Dica C (2006) Effects of gasoline-air enrichment with HRG gas on efficiency and emissions of a SI engine. SAE Technical Paper
El Kady M et al (2020) Parametric study and experimental investigation of hydroxy (HHO) production using dry cell. Fuel 282:118825
El Soly A et al (2021) Comparative experimental investigation of oxyhydrogen (HHO) production rate using dry and wet cells. Int J Hydrogen Energy 46(24):12639–12653
Gad M, Razek SA (2021) Impact of HHO produced from dry and wet cell electrolyzers on diesel engine performance, emissions and combustion characteristics. Int J Hydrogen Energy 46(43):22277–22291
Gohar GA, Hassan R (2017) Comparative analysis of performance characteristics of CI Engine with and without HHO gas (Brown Gas). Adv Automob Eng 6(172):2
Hadi MS et al (2019) Hydrogen embrittlement of 316L stainless steels exposed in 1.0 M hydrochloric acid solution. Ann Chim Sci Mater 43(6):369–375
Ismail TM et al (2018) Performance of hybrid compression ignition engine using hydroxy (HHO) from dry cell. Energy Convers Manage 155:287–300
Manu P, Sunil A, Jayaraj S (2016) Experimental investigation using an on-board dry cell electrolyzer in a CI engine working on dual fuel mode. Energy Procedia 90:209–216
Paparao J, Murugan S (2021) Oxy-hydrogen gas as an alternative fuel for heat and power generation applications—a review. Int J Hydrogen Energy 46(76):37705–37735
Saravanan N et al (2008a) An experimental investigation on hydrogen as a dual fuel for diesel engine system with exhaust gas recirculation technique. Renew Energy 33(3):422–427
Saravanan N et al (2008b) Combustion analysis on a DI diesel engine with hydrogen in dual fuel mode. Fuel 87(17–18):3591–3599
Sharma PK et al (2020) Characterization of the hydroxy fueled compression ignition engine under dual fuel mode: experimental and numerical simulation. Int J Hydrogen Energy 45(15):8067–8081
Subramanian B, Ismail S (2018) Production and use of HHO gas in IC engines. Int J Hydrogen Energy 43(14):7140–7154
Subramanian B, Thangavel V (2020a) Analysis of onsite HHO gas generation system. Int J Hydrogen Energy 45(28):14218–14231
Subramanian B, Thangavel V (2020b) Experimental investigations on performance, emission and combustion characteristics of Diesel-Hydrogen and Diesel-HHO gas in a Dual fuel CI engine. Int J Hydrogen Energy 45(46):25479–25492
Wang H-K et al (2012) Effect of regulated harmful matters from a heavy-duty diesel engine by H_2/O_2 addition to the combustion chamber. Fuel 93:524–527
Yadav VS, Soni S, Sharma D (2014) Engine performance of optimized hydrogen-fueled direct injection engine. Energy 65:116–122
Yadav Milind S et al (2011) Investigations on generation methods for oxy-hydrogen gas, its blending with conventional fuels and effect on the performance of internal combustion engine. J Mech Eng Res 3(9):325–332
Yilmaz AC, Uludamar E, Aydin K (2010) Effect of hydroxy (HHO) gas addition on performance and exhaust emissions in compression ignition engines. Int J Hydrogen Energy 35(20):11366–11372 (2010)

Chapter 7
E-Fuels and Electric Vehicles for Decarbonizing Transport Sector: A Comparative Analysis

Durgesh Srivastav, Nagesh D. Patil, and Pravesh Chandra Shukla

Abstract The transport sector is one of the major sources of greenhouse gas emissions, and efforts are being put to control these emissions worldwide. The present topic includes the comparison of technological, economic, and environmental aspects of e-fuels and electric vehicles (EVs) toward decarbonization. E-fuels are promising in terms of overall CO_2 emission reduction to the environment as it can be produced by capturing CO_2 from the combustion sources or ambient. While EVs are important in the context of urban pollution due to no tailpipe exhaust emissions on road conditions. The chapter discusses these two alternatives for their possible adaptation by considering factors like production cost, implementation, government policies, application area, etc. Comparative discussion for the adaptation of e-fuels and EVs are with a perspective of their advantages in terms of decarbonization of transport sector.

Keywords E-fuels · Electric vehicles · Hydrogen · Carbon dioxide · Renewable energy

7.1 Introduction

Transportation sector is the prime contributor to greenhouse gas (GHG) emissions, accounting for approximately 24% of carbon dioxide (CO_2) emissions resulting from fuel combustion (Solaymani 2019). Reduction in CO_2 emissions from transportation sector is an viable topic of discussion. Adoptation of Electric Vehicles (EVs) and

D. Srivastav (✉)
Discipline of Electric Vehicle Technology, Department of Electrical Engineering, Indian Institute of Technology Bhilai, Durg, India
e-mail: durgeshsr@iitbhilai.ac.in

N. D. Patil · P. C. Shukla
Department of Mechanical Engineering, Indian Institute of Technology Bhilai, Durg, India

N. Sharma et al. (eds.), *Sustainable, Scalable and Storable E-Fuels for Decarbonising Transport Sector*, Energy, Environment, and Sustainability,
https://doi.org/10.1007/978-981-96-7388-9_7

promoting e-fuels utilization is a potential way to reduce overall global CO_2 reduction (Yu et al. 2024). Internal combustion (IC) engine, initially created by Étienne Lenoir in 1860 and later improved by Nikolaus Otto in 1876 (Jonkers 2019), brought significant change in transportation. It provided a powerful and efficient means of converting chemical energy into mechanical work. IC engines have become the essential element of modern transportation, providing power to cars, lorries, ships, and airplanes. Improved engine efficiency, higher power density and compactness made it a preferred choice over steam engines (Stone 1999). The reliance of combustion engines on fossil fuels has led to environmental issues of air pollution and the release of greenhouse gases, which is significantly affecting both human health and the ecosystem (Reitz et al. 2020). EVs, mostly regarded as a modern innovation, have a history of development that can be traced back to the nineteenth century. A Scottish inventor Robert Davidson built the first electric car in 1837 which was propelled by galvanic cells. His battery based locomotive was displayed at the Royal Scottish Society of Arts Exhibition in 1841 (Kanna et al. 2018). During the late nineteenth and early twentieth centuries, electric cars were widely present. During the early 1900s, almost one-third of all automobiles on American highways were electric cars (Department of Energy, Government of United States of America). The rise in popularity of gasoline-powered cars was mostly due to its ability to provide longer driving ranges and faster refueling times.

E-fuels, are synthetic fuels produced from green hydrogen and CO_2 captured from atmosphere using electricity produced from renewable resources, is a relatively new scientific advancement. It is important to understand the difference between the synthetic fuels and e-fuels. Primarily, the process of fuel-making is same in both the cases while e-fuels differs in such a way that the energy required in the production of e-fuels are green energy (electricity generated from renewable resources) while in case of production of synthetic fuel the energy is supplied from conventional energy resources. The major e-fuels are e-methane, e-methanol and ammonia, hydrogen (H_2) produced from renewable energy is reacted with CO_2 which is captured from the atmosphere using renewable energy resources to produce e-methane, similarly e-methanol and ammonia is produced by reacting H_2 with appropriate chemical compound (Colelli et al. 2023). Recently, the concept of e-fuels has been increasingly popular as a means to neutralize carbon emissions from conventional combustion engines, without requiring a complete overhaul of the current transportation infrastructure. The development of e-fuel is significantly influenced by advancements in renewable energy and electrolysis technologies.

The twenty-first century has experienced a revival in EVs, mostly as a result of progress in battery technology, increased environmental awareness, and favorable government policies toward EVs. Modern EVs provide numerous advantages, including no tail pipe emissions, lower operational costs, and less mechanical part complexity (Costa et al. 2021). The International Energy Agency (IEA) estimated in 2023 that the electric passenger car's global stock surpassed 14 million, the electric car sales increased from 4% in 2020 to 18% in 2023 worldwide and sales continue to grow rapidly (International Energy Agency). Modern EVs are powered by Li-ion batteries consisting of small Li-ion cells. Advancements in battery technology are a major driving force behind the EV revolution. Li-ion cells were first introduced to the market in 1991. Li-ion cell made significant progress in terms of their ability to store energy, reduced cost, and fast charging and dominated the market (Blomgren 2016). Technological advancements have enhanced EVs' competitiveness compared to conventional internal combustion engine (IC) vehicles. The price of Li-ion batteries has experienced a significant decrease, approximately 14 % drop from 2022 to 2023 (BloombergNEF, November 26th 2023). The decrease in cost, combined with advancements in energy density, has rendered EVs more attainable and convenient for daily usage. In addition, the ongoing progress in battery recycling and second-life applications is effectively resolving problems related to the environmental impact of battery production and disposal (Zhao et al. 2021). The implementation favorable rules for EVs, along with the increasing environmental consciousness among customers, has developed a favorable atmosphere for the expansion of EVs. E-fuels offer a supplementary alternative for decarbonizing carbon emissions in the transportation sector, particularly in regions where electrification is difficult, such as long-haul trucks, aviation, and shipping. E-fuels can function as a direct replacement for conventional fossil fuels by utilizing current infrastructure and IC engines, so aiding in a seamless shift toward a low-carbon future. E-fuel manufacturing requires the capturing of CO_2 from the atmosphere and industrial origins. This CO_2 is then mixed and react with hydrogen in the presence of suitable catalyst to produce e-fuels. E-fuels can potentially achieve carbon neutrality by compensating for the CO_2 released during combustion with the CO_2 absorbed during production (Reddy et al. 2023). In the future, e-fuels could become more competitive due to technological breakthroughs in renewable energy, advanced electrolysis techniques and favorable implementation policies for the production of e-fuel at mass level.

This chapter aims to offer a thorough comparative discussion on the possible impact of EVs and e-fuels on the decarbonization of transport industry contributing to a sustainable and environmentally friendly future with reduced carbon emissions. The key aspect highlighted in this chapter are discussion about production process of e-fuels and its advantages and challenges, the prospective of EVs, brief comparison about EV and IC engine vehicles, classifications of EVs including challenges and benefits, technological development of EVs, and lastly there is comparative study is performed for both EV and e-fuels by taking into account for production cost, implementation, government policies, application area, and drawbacks. Both EVs and e-fuels possess distinct benefits and challenges. It is important to analyze and compare the prospects of EVs and e-fuels for a sustainable and carbon zero approach in transportation sector.

7.2 E-Fuels for Transportation Sector

E-fuels, also referred to as electro-fuels, are a recent technological innovation in the field of renewable fuels. These fuels are specifically designed to achieve carbon neutrality in the transportation industry. These fuels are mostly the result of chemical processes involving captured CO_2 from the atmosphere and hydrogen created by utilizing sustainable green energy sources producing e-methane (CH_4).

$$CO_2 + 2H_2 \rightarrow CH_4 + O_2$$

In the similar fashion, e-methanol (CH_3OH) and e-ammonia (NH_3) can also be produced by using captured carbon and nitrogen from the atmosphere and reacted with the green hydrogen

$$CO_2 + 3H_2 \rightarrow CH_3OH + H_2O$$

$$N_2 + 3H_2 \rightarrow 2NH_3$$

Figure 7.1 depicts the production process of e-fuels, which involves two primary components: the production of hydrogen and the capture of carbon. Both of these aspects are considered to be important. In order to maintain zero emissions, green hydrogen is preferred choice for the production of e-fuels. During production of hydrogen, the electricity that is required for the production of hydrogen is preferred to be supplied by solar panels or wind power, and other green sources of electricity. Hydrogen is produced through the electrolysis process. Concerning to the capture of CO_2, number of different technologies can be utilized to capture carbon, discussed in later part of this chapter. After hydrogen has been produced and carbon has been captured, the next step is to begin the process of fuel synthesis. E-methane (CH_4) and e-methanol (CH_3OH) are those that are produced as a result of the reaction between carbon and hydrogen. Ammonia is produced as a result of the reaction between hydrogen and nitrogen, which is taken from the environment.

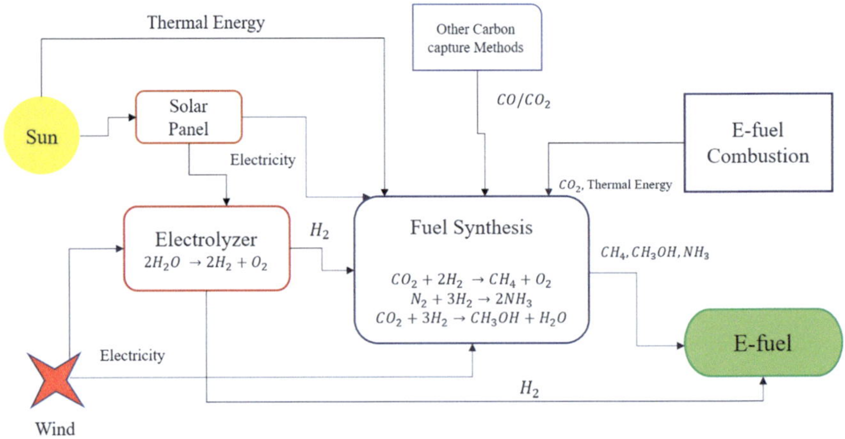

Fig. 7.1 E-Fuel production process (Huey et al. 2024)

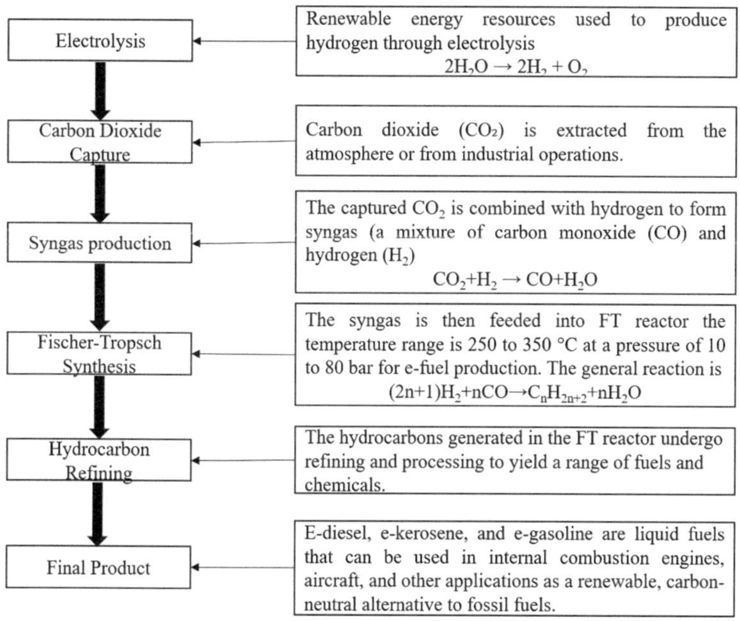

Fig. 7.2 Steps involved in the production of e-fuel (Chen et al. 2021)

7.2.1 Understanding the Steps Involved in E-Fuel Production

Figure 7.2 represents the steps involved in the production of e-fuel, namely electrolysis, CO_2 capture, syngas manufacturing, Fischer–Tropsch synthesis, and hydrogen production. The Fisher Tropsch reaction is a crucial reaction mechanism employed in this procedure. The Fischer–Tropsch synthesis (FTS) is a catalytic process that converts syngas, which is a mixture of carbon monoxide (CO) and hydrogen (H_2), into a substance like petroleum known as FT crude. The FT crude may be easily transformed into various liquid hydrocarbons that are suitable for existing fleet of IC engines. The Fischer–Tropsch process was initially developed during the 1920s and subsequently named for its discoverers, Franz Fischer and Hans Tropsch (Klerk 2000). The FT process has the ability to convert several types of raw materials, which can be provided in the form of CO and H_2 components into useful e-fuel.

7.2.2 Large Scale Hydrogen Production Technology

The primary objective of large-scale green hydrogen production technology is to generate hydrogen using environmentally sustainable ways, with a particular emphasis on utilizing renewable energy sources. Electrolysis is the primary technique employed to separate water into hydrogen and oxygen by utilizing electricity derived from renewable sources such as wind, solar, or hydroelectric power. This process does not release any greenhouse gases, making it a sustainable alternative to conventional techniques of hydrogen synthesis such as steam methane reforming. Advancements in electrolyzer technology, such as the utilization of polymer electrolyte membrane (PEM) and solid oxide electrolyzers, are improving efficiency and decreasing expenses (Viscardi et al. 2021). Furthermore, the progress in incorporating renewable energy, storing energy, and managing the power grid is essential for expanding the production of green hydrogen (Table 7.1).

Table 7.1 Green hydrogen production technology (Singh et al. 2022)

Technology		Description	Advantages	Disadvantages	TRL	Cost ($ per kg H_2)
Electrolysis	Alkaline	Through the utilization of electricity as the source of energy, the electrolysis process is able to directly transform water into hydrogen and oxygen. The primary components of electrolysis are an cathode and a anode, which are separated from one another by an electrolyte	It was the first water splitting technology ever developed, and it is quite inexpensive	Corrosive liquid electrolyte	9	2.30 (with $0.037 per kwh as the electricity cost)
	Solid-oxide		Can leverage both heat and electrical energy	Have a high temperature requirement. It takes longer to start up	5	2.30 (with $0.037 per kwh as the electricity cost)
	Polymer electrolyte (or proton exchange) membrane (PEM)		Has the ability to function at high current densities which leads to faster response times	High-priced components that contribute to increase in the overall cost	6–8	2.30 (with $0.037 per kwh as the electricity cost)
Naturally occurring free-state H_2 found in geological media			Oil and gas drilling technology that is already in use can be used to extract it	A lack of understanding exists on the geology of its occurrence	1	Most economical

7.2.3 *Carbon Capture Technology*

Carbon capture is a crucial component in the production of E-fuels. CO_2 undergoes a reaction with hydrogen gas (H_2) to form syngas, which is a combination of water (H_2O) and carbon monoxide (CO). The carbon monoxide (CO) is subsequently employed in the FT Reactor to generate e-fuels. Several ways are available for extracting CO_2, Table 7.2 illustrates several popular CO_2 capture technologies.

7.2.4 *Benefit and Challenges of E-Fuels*

E-fuels are emerging technology in the field of automobile, there are various benefits like they can be used with the existing fleet if IC engine vehicles and fuel stations, it can provide energy security and better storage ability as well as challenges like high production cost, limited efficiency and infrastructural requirements (Viscardi et al. 2021). In Table 7.3 benefits and challenges of e-fuels are discussed briefly.

7.3 Prospects of Electric Vehicles (EVs)

EVs are those vehicles which are powered by battery and the wheel power is obtained with the help of electric motor. The battery which is preferred for EVs are Li-ion battery, it is because of better energy density of Li-ion battery as compared to lead acid battery, higher life and compact in nature (Yudhistira et al. 2022). Figure 7.3 is illustrating a simple EV system it shows all the important components the Battery pack, Power converter, Electric motor, Transmission system, and wheels along with Battery thermal management system (BTMS) and Battery management system (BMS) as well as charger. The electric motor is powered with the help of Li-ion battery and with the help of transmission system the power is transferred to the wheel axle to propel the vehicle (Un-Noor et al. 2017).

EVs are getting popular these days because of its help in reducing pollution from urban areas and also it reduces the dependency on the fossil fuels. The emissions due to electric vehicle is not absolute zero, the battery of EV is getting charged by the electricity and that electricity production is emitting the pollutants. The major difference in the emission from IC engine vehicle and electric vehicle is, the emission is continuously from tailpipe during operation while EVs are a zero-tailpipe emission vehicle. Emissions from EV is indirect and normally happens at the location where electricity generation occurs, i.e., power plants, electric vehicle manufacturing units, battery manufacturing plants, and OEMs of EVs. Figure 7.4 is illustrating a comparison between EVs and ICs (Poornesh et al. 2020).

There are various kind of EVs are available in the market like battery electric vehicles (BEV) and hybrid electric vehicles (HEV). BEVs are those vehicles which

Table 7.2 Major carbon capture technologies (Singh et al. 2022)

Characteristics	Pre-combustion	Post combustion				
		Traditional method	Specialized methods			
			Cryogenic	Oxyfuel combustion	Chemical looping	
Source of CO_2 emission	The syngas coming from gasification equipment before being used for combustion	The flue gas which is produced from the combustion process involving air and fuel	The flue gas which is produced from the combustion process involving air and fuel	The flue gas which is produced from the combustion process involving oxygen and fuel	The flue gas which is produced from the combustion process	
Capture mechanism	Sorption with variations in temperature and/or pressure	Sorption with variations in temperature and/or pressure	Direct transition from the gaseous state to either the liquid or solid state	Sorption with variations in temperature and/or pressure	Sorption with variations in temperature and/or pressure	
CO_2 capturing agents	Membranes and solvents	Adsorbents, Solvents and membranes	Cryogenic cooling using liquid solvents	Membranes and solvents	Membranes and solvents	
Outputs	Carbon dioxide (CO_2) and Hydrogen (H_2)	Carbon dioxide (CO_2) and Nitrogen (N_2)	CO_2, N_2, SO_x, NO_x, and Hg	CO_2 and Steam	CO_2 and O_2	
Advantages	Carbon Capture efficiency is high	Most developed technology	A significant amount of carbon dioxide (CO_2) is captured	Results in a significant accumulation of carbon dioxide in the flue gas	An economically efficient alternative to oxy-combustion, with comparable advantages	
Disadvantages	Immature technology	Carbon Capture efficiency is less	Relative high capital investment or expenses (CAPEX)	Relative high capital investment or expenses (CAPEX)	Immature technology	

Table 7.3 Benefits and challenges of e-fuels

S.No	Benefits	Challenges
1	*Carbon neutrality*: E-fuels, which is produced from renewable source of energy, achieve carbon neutrality when utilized in automobiles or other uses	*Energy intensive production*: The manufacture of e-fuels necessitates substantial energy inputs, which may be sourced from non-renewable resources in the absence of renewable alternatives
2	*Compatibility*: E-fuels are compatible with current IC engines and infrastructure, requiring no significant adjustments, unlike certain other alternative fuels	*Cost*: At now, the production of e-fuels is typically more costly than that of fossil fuels, mostly because to the energy-intensive production methods and low economies of scale
3	*Energy security*: E-fuels have the potential to decrease reliance on fossil fuels and improve energy security by expanding the range of available fuel sources	*Limited efficiency*: The conversion efficiency of renewable energy into E-fuels, such as hydrogen or synthetic fuels, is comparatively lower than the direct use of renewable electricity in applications such as EVs
4	*Storage stability*: Unlike to certain renewable energy sources such as electricity, e-fuels have the advantage of being able to be stored for long time without experiencing significant energy loss or deterioration	*Environmental impact*: E-fuels might potentially have environmental consequences, such as carbon emissions from non-renewable energy sources or land-use effects for biomass-based fuels, depending on the production processes and energy sources employed
5	*Global application*: E-fuels have the potential to be manufactured and utilized worldwide, which might help decrease the reliance on particular energy sources in specific regions	*Infrastructure requirements*: The manufacturing, storage, and delivery of e-fuels may necessitate substantial expenditures in infrastructure, thus presenting obstacles to their general acceptance

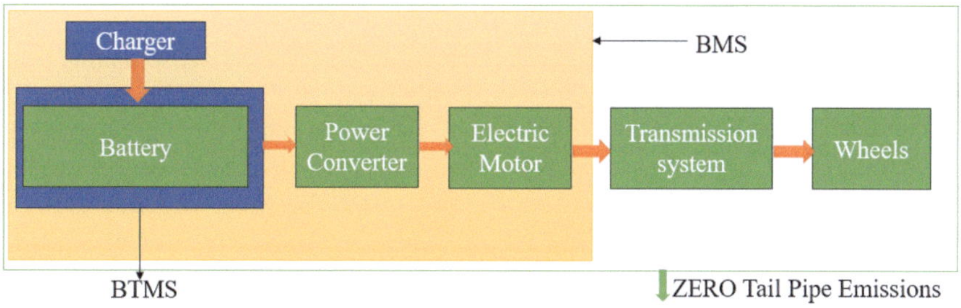

Fig. 7.3 A simple EV system

are only powered by the electric motor and battery and they only charged with the electric mains while the HEVs are those vehicles which are powered by both the IC engine and the electric motor and they are optimized for the best efficiency. Plugin hybrid (PHEV) is those which are both can be fueled with fossil fuel and also charged with electricity by plugin into charging socket, and fuel cell electric vehicle (FCEV)

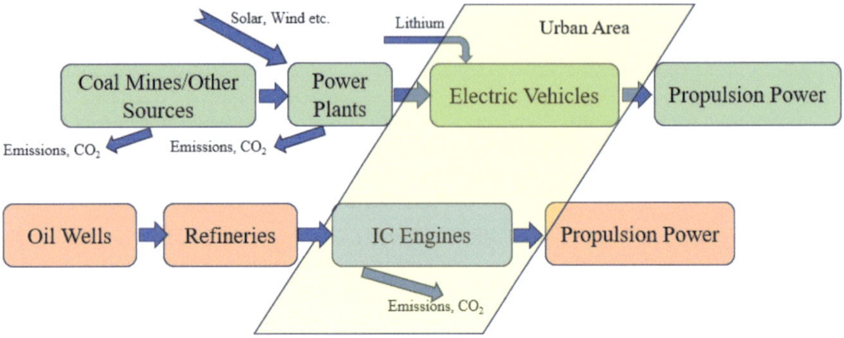

Fig. 7.4 Internal combustion engines Vs electric vehicles

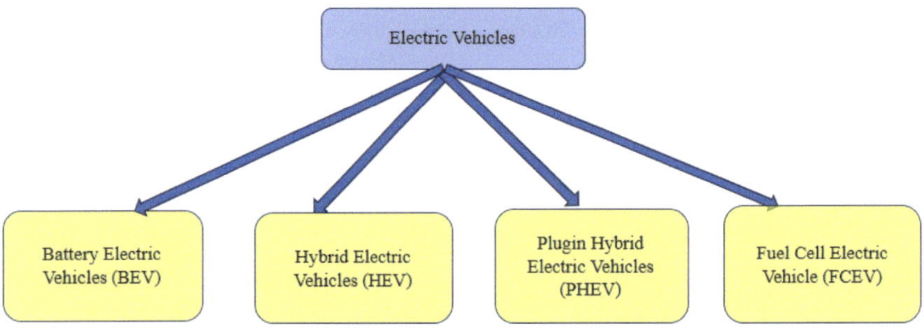

Fig. 7.5 Broader classification of electric vehicle (e-AMRIT, NITI Ayog, Government of India)

are those which are zero emission vehicle the electricity is provided by fuel cell technology. Figure 7.5 is showing the classification of EV.

From the future perspective the EVs are in advantageous side as compared to e-fuels technology the sales of EV are growing rapidly. In 2023, the global registration of new EVs reached to 14 million, altogether the overall number of EVs on roads to 40 million. In 2023, the sales of EV saw a substantial increase of 3.5 million units compared to the previous year, indicating a year-on-year growth rate of 35%. In 2023, electric cars were around 18% of total car sales, a significant increase from 14% in 2022. These data suggest that the growth of electric car markets continues to be strong as they get more developed. In 2023, battery electric cars comprised 70% of the total electric car inventory (Alanazi 2023).

7.3.1 Benefit and Challenges of Electric Vehicle

There are various benefits of electric vehicle like zero-tailpipe emission, lower operating cost, less reliance on fossil fuels, etc. However, despite the advantages, there are still several challenges are also associated with the EVs like Improper charging infrastructure, Range anxiety, High initial cost, Battery thermal runaway of EVs, etc. (Alanazi 2023). Table 7.4 is briefly presenting the Benefits and challenges associated with EVs.

7.3.2 Technological Advancement in EVs

Since their introduction in the nineteenth century, EVs have experienced substantial technological advancements. The journey started in 1837 with Robert Davidson's electric-powered road vehicle, which utilized non-rechargeable batteries. Gaston Planté's invention of the rechargeable lead-acid battery in 1859 (Kurzweil 2010), leads to development of practical EVs, including Gustave Trouvé's electric tricycle, which was exhibited in Paris in 1881 (Desmond 2018). In the 1892, William Morrison's electric carriage was a significant milestone in the commercial adoption of EVs, which had acquired traction by the late nineteenth century (Anderson 2010). EVs were approximately one-third of all vehicles in the United States at the turn of the nineteenth century, owing to their quiet and dependable operation. However, their popularity experienced a rapid decline following the introduction of Ford's affordable gasoline-powered Model T in 1908 and the development of the electric starter in 1912. The oil crises of the 1970s stimulated renewed interest in alternative propulsion technologies, such as electric vehicle (Department of Energy, Government of United States of America). The introduction of lithium-ion (Li-ion) batteries in the 1991 (Wakihara 2001) revolutionized EV technology by providing increased energy density and extended lifespans. The 2008 Tesla Roadster marked the beginning of the industry's transformation by showcasing EVs as high-performance vehicles. Models such as the Nissan Leaf and Tesla Model S achieved mass-market appeal in the 2010s as a result of advancements in affordability, charging infrastructure, and range. This marked a turning point. Subsidies and regulatory incentives were implemented by governments worldwide, resulting in an exponential increase in the sales of EVs. Advancements in solid-state batteries, rapid charging networks, and vehicle-to-grid (V2G) integration were the primary factors contributing to the rise in EV sales (BloombergNEF, June 8th 2023).

In recent years, the EVs industry has prioritized the development of sustainable materials for manufacturing, Effective battery management system, Efficient battery thermal management system, and wireless charging. As of 2024, EVs are cost-competitive with internal combustion engines (ICEs) (Sesterhenn 2023), have a range approximately 400 km. The evolution of EVs from experimental novelties to essential components of sustainable transportation is emphasized by the transition toward electrification, which is supported by climate policies and consumer demand.

Table 7.4 Benefits and challenges of electric vehicle operation

S.No	Benefits	Challenges
1	*Environmental benefits*: EVs emit no pollutants from their tailpipes, so contributing to the reduction of both greenhouse gas emissions and air pollution. EVs, when powered by renewable energy sources, have the capacity to substantially decrease the amount of carbon emissions released into the environment	*Limited range*: Even though they are getting better, EVs usually have a shorter range than gas-powered cars. For long trips, might need to plan carefully
2	*Economic benefits*: Electricity is typically more cost-effective than petrol or diesel. EVs also possess lesser moving parts as compared to conventional ICE vehicle resulting in decreased expenses for maintenance. In addition, numerous governments provide, tax credits, incentives, and rebates to encourage the purchase of EVs	*Charging infrastructure*: The situation of charging station is not very well at present stage. In the urban areas the charging stations are present in Supermarket, Malls, Gas Stations, and Hospitals while in the rural areas and on highways the number of charging stations is not sufficient. Government as well as private players need to develop the charging infrastructure as fastly as possible
3	*Performance benefits*: Electric motors deliver immediate torque, resulting in rapid and smooth acceleration. In addition, EVs produce less noise pollution compared to conventional internal combustion engine vehicles	*High initial cost*: The initial cost of an EV while purchasing is high as compared to same model ICE vehicle the reason behind this is the higher cost of the Li-ion batteries which is utilized in EVs but as the battery technology advancing the cost of Li-ion battery is going down and once the Li-ion batteries developed at affordable cost the initial cost of EVs will be reduced
4	*Energy independence*: EVs can enhance energy security by reducing reliance on imported oil through the use of electricity as a fuel	*Battery Life and Recycling*: The major challenge with the EV Li-ion battery is they degrade with time and they have limited life in general Li-ion batteries have around 1000 Cycle life it depends on the chemistry of Li-ion cells used in the battery pack. Furthermore, once the Li-ion battery is dead the recycling is also a challenge. Researchers are working to improve the recycling methodology of Li-ion cells
5	*Technological advancements*: EVs comes with cutting-edge technology, such as sophisticated driver assistance systems, seamless connection, and state-of-the-art entertainment systems	*Battery thermal runaway*: Battery thermal runaway of an EV battery is a major challenge. When Li-ion cells work (charge and discharge) they produces significant amount of heat due to joules heating and chemical (entropic) heating. The normal battery operating temperature range for Li-ion battery is 15–45 °C (Velumani and Bansal 2022). Researchers are working to improve the cooling efficiency of EV Battery pack to make it safer and more durable

7.4 Comparative Analysis of E-fuels and Electric Vehicles (EVs)

Both the electric vehicle and E-fuels can help in the decarbonize the transport sector in their capabilities. EVs can help in decarbonizing the urban areas and reduce the emissions while the use of e-fuels can help in overall transportation sector decarbonization provided the H_2 produced and CO_2 captured using green energy. In Table 7.5 a comparative study is presented from different aspect for e-fuels and electric vehicle.

Table 7.5 E-fuels and EVs comparative analysis

Aspect	E-fuels	Electric vehicles (EVs)
Production cost	1. High CO_2 capture cost and H_2 production	1. High cost of Li-ion battery production
Implementation	1. Compatible with existing ICE vehicles 2. Can use current fuel distribution infrastructure 3. Scaling up production capacity is challenging	1. Requires significant investment in charging infrastructure 2. Requires upgrades to the power grid and integration of renewable energy 3. Advances in battery technology are addressing range and charging time concerns
Government Policies	1. Government and policy makers need to make policies regarding e-fuel production 2. Need incentives and rebates for e-fuel producers	1. Incentives for EV Purchase, Tax rebate, and road tax exemption are already available for EVs in many countries in India FAME*1 and FAME*2 is implemented regarding this 2. Support for R&D in battery technology, battery thermal management system, fast charging, exhausted battery recycling, and charging infrastructure
Application Areas	1. Suitable for aviation and shipping 2. Applicable for heavy-duty transport	1. Suitable for personal transportation 2. Compatible for urban areas and public transportation
Drawbacks	1. Less energy efficient in comparison to direct electrification 2. The absence of financial supporting and technological advancements leads to expensive production expenses	1. Reliance on limited resources for the development of batteries 2. Efficient battery recycling is necessary to reduce the environmental consequences 3. Establishing the required charging infrastructure necessitates significant financial commitment and cooperation

* Faster Adoption and Manufacturing of Electric Vehicles (FAME)—Indian Government policy for the adaption and manufacturing of EV

7.5 Summary

E-fuels having the benefit of being compatible with current IC engine based vehicles and fueling infrastructure, which could make the transition to low-carbon transport sector easier. However, their manufacturing process requires a significant amount of energy and is less effective when compared to direct electrification in the present scenario. In the other hand EVs, offer superior energy efficiency and produce no emissions from their tailpipes. However, they necessitate substantial expenditures in charging infrastructure and breakthroughs in battery technology to alleviate concerns about, range anxiety, charging infrastructure and battery thermals control. The study's findings suggest that e-fuels have the potential to contribute to the reduction of carbon emissions in challenging sectors such as aviation and shipping. Although, production of e-fuels requires more government support, rebates and tax exemption. However, EVs are deemed more efficient for light-duty cars, as they have higher energy efficiency and lower emissions throughout their lifespan. An integrated strategy, i.e., using both the e-fuels and EVs in compatible application area is advised to accomplish a thorough and effective reduction of carbon emissions in the transport industry. Furthermore, this book chapter highlights the individual benefits and challenges of both the Electric vehicle and e-fuels for transportation sectors.

References

Alanazi F (2023) Electric vehicles: benefits, challenges, and potential solutions for widespread adaptation. Appl Sci 13(10):6016

Anderson CD, Anderson J (2010) Electric and hybrid cars: a history. McFarland

Blomgren GE (2016) The development and future of lithium ion batteries. J Electrochem Soc 164(1):A5019

BloombergNEF (November 26th, 2023) Press release titled 'Lithium-ion battery pack prices hit record low of $139/kWh'. https://about.bnef.com/insights/clean-energy/lithium-ion-battery-pack-prices-hit-record-low-of-139-kwh/. Published on November 26th, 2023 and Accessed on 20th August 2024

BloombergNEF (June 8th, 2023) Press release titled 'Electric vehicle fleet set to hit 100 million by 2026, but stronger push needed to stay on track for net zero'. https://about.bnef.com/insights/clean-transport/electric-vehicle-fleet-set-to-hit-100-million-by-2026-but-stronger-push-needed-to-stay-on-track-for-net-zero/. Published on June 8th, 2023 and accessed on 20th August 2024

Chen Y, J, Duyar MS, Ordomsky VV, Khodakov AY, Liu J (2021) Carbon-based catalysts for Fischer-Tropsch synthesis. Chem Soc Rev 50(4):2337–2366

Colelli L, Segneri V, Bassano C, Vilardi G (2023) E-fuels, technical and economic analysis of the production of synthetic kerosene precursor as sustainable aviation fuel. Energy Convers Manage 288:117165

Costa CM, Barbosa JC, Castro H, Gonçalves R, Lanceros-Méndez S (2021) Electric vehicles: to what extent are environmentally friendly and cost effective?—comparative study by European countries. Renew Sustain Energy Rev 151:111548

Department of Energy, Government of United States of America, https://www.energy.gov/articles/history-electric-car. Accessed on 20th August 2024

Desmond K (2018) Electric motorcycles and bicycles: a history including scooters, tricycles, Segways and monocycles. McFarland

e-AMRIT, NITI Ayog, Government of India, https://e-amrit.niti.gov.in/types-of-electric-vehicles. Accessed on 20th August 2024

Huey C, Metghalchi H, Levendis Y (2024) E-fuels as reduced carbon emission options. ASME Open J Eng

International Energy Agency, https://www.iea.org/energy-system/transport/electric-vehicles. Accessed on 20th August 2024

International Energy Agency, https://www.iea.org/reports/global-ev-outlook-2024. Accessed on 20th August 2024

Jonkers HR (2019) Advanced internal combustion engine sensor analysis by means of modelling and real-time measurement. Bachelor's thesis, University of Twente

Kanna GRR, Muthulakshmi K, Sasiraja RM (2018) History of electric vehicle. J Electr Eng (2018)

de Klerk A (2000) Fischer–tropsch process. Kirk-Othmer encyclopedia of chemical technology 1–20

Kurzweil P (2010) Gaston Planté and his invention of the lead–acid battery—the genesis of the first practical rechargeable battery. J Power Sources 195(14):4424–4434

Poornesh K, Nivya KP, Sireesha K (2020) A comparative study on electric vehicle and internal combustion engine vehicles. In: 2020 international conference on smart electronics and communication (ICOSEC). IEEE, pp 1179–1183

Reddy VJ, Hariram NP, Maity R, Ghazali MF, Kumarasamy S (2023) Sustainable e-fuels: green hydrogen, methanol and ammonia for carbon-neutral transportation. World Electr Veh J 14(12):349

Reitz RD, Ogawa H, Payri R, Fansler T, Kokjohn S, Moriyoshi Y, Agarwal AK et al (2020) IJER editorial: the future of the internal combustion engine. Int J Engine Res 21(1):3–10

Sesterhenn M (2023) Electric engines as an alternative powertrain solution in the urban freight logistics industry

Singh H, Li C, Cheng P, Wang X, Liu Q (2022) A critical review of technologies, costs, and projects for production of carbon-neutral liquid e-fuels from hydrogen and captured CO_2. Energy Adv 1(9):580–605

Solaymani S (2019) CO2 emissions patterns in 7 top carbon emitter economies: the case of transport sector. Energy 168:989–1001

Stone R (1999) Introduction to internal combustion engines, vol 3. Macmillan, London

Un-Noor F, Padmanaban SK, Mihet-Popa L, Mollah MN, Hossain E (2017) A comprehensive study of key electric vehicle (EV) components, technologies, challenges, impacts, and future direction of development. Energies 10(8):1217

Velumani D, Bansal A (2022) Thermal behavior of lithium-and sodium-ion batteries: a review on heat generation, battery degradation, thermal runway–perspective and future directions. Energy Fuels 36(23):14000–14029

Viscardi R, Bassano C, Nigliaccio G, Deiana P (2021) The potential of E-fuels as future fuels. ENEA Energ Ambiente Innov 1:112–116

Wakihara M (2001) Recent developments in lithium ion batteries. Mater Sci Eng R Rep 33(4):109–134

Yu X, LeBlanc S, Sandhu N, Wang L, Wang M, Zheng M (2024) Decarbonization potential of future sustainable propulsion—a review of road transportation. Energy Sci Eng 12(2):438–455

Yudhistira R, Khatiwada D, Sanchez F (2022) A comparative life cycle assessment of lithium-ion and lead-acid batteries for grid energy storage. J Clean Prod 358:131999

Zhao Y, Pohl O, Bhatt AI, Collis GE, Mahon PJ, Rüther T, Hollenkamp AF (2021) A review on battery market trends, second-life reuse, and recycling. Sustain Chem 2(1):167–205

Chapter 8
Integrated Power Electronics Solutions for Sustainable, Scalable, and Storable E-Fuel Production in Decarbonizing Transport

Vivek Kumar, Padmini Singh, Jai Gopal Gupta, and Varun Gupta

Abstract The urgent need to decarbonize the transport sector has catalyzed the development of sustainable alternatives, including e-fuels synthesized from renewable energy sources. This chapter investigates the integration of advanced power electronics solutions to enable scalable, sustainable, and storable e-fuel production. Emphasizing efficiency and reliability, it explores the pivotal role of power electronics in hydrogen production via electrolysis, energy management, and renewable energy integration. Case studies and practical applications demonstrate the transformative potential of these technologies in addressing challenges such as renewable energy variability, production scalability, and storage efficiency. The chapter provides insights into creating a resilient transport infrastructure while minimizing greenhouse gas emissions by detailing innovative strategies, including modular systems and AI-driven controls. It is a comprehensive guide for researchers, policymakers, and industry professionals committed to advancing decarbonized transport solutions.

Keywords Carbon emissions · Energy storage · Electrolysis efficiency · E-fuels · Power electronics · Renewable energy · Transport sustainability

V. Kumar
Electrical and Electronics Engineering Department, National Institute of Technology Sikkim, Ravangla, India
e-mail: vivekkumar@nitsikkim.ac.in

P. Singh
Electrical Engineering Department, Indian Institute of Technology Jammu, Jammu, India
e-mail: padmini.singh@iitjammu.ac.in

J. G. Gupta (✉)
Mechanical Engineering Department, National Institute of Technology Sikkim, Ravangla, India
e-mail: jaig@nitsikkim.ac.in

V. Gupta
Electronics and Communication Engineering Department, National Institute of Technology Sikkim, Ravangla, India
e-mail: varungupta@nitsikkim.ac.in

N. Sharma et al. (eds.), *Sustainable, Scalable and Storable E-Fuels for Decarbonising Transport Sector*, Energy, Environment, and Sustainability,
https://doi.org/10.1007/978-981-96-7388-9_8

8.1 Introduction

The transport sector substantially contributes to global carbon emissions, necessitating innovative approaches to mitigate its environmental impact. E-fuels, synthesized from renewable energy, offer a promising solution due to their compatibility with existing internal combustion engines and infrastructure. However, their production poses efficiency, scalability, and storage challenges. Integrated power electronics solutions are instrumental in overcoming these challenges by optimizing production processes and ensuring energy supply stability. The urgent need to reduce carbon emissions has driven significant interest in e-fuels, which can serve as drop-in replacements for conventional fossil fuels. E-fuels, such as synthetic gasoline, diesel, and jet fuel, combine hydrogen from water electrolysis with carbon dioxide from the air or industrial processes. This synthesis requires substantial amounts of renewable energy, and the efficiency of converting this energy into e-fuels is paramount to making the process viable. Advanced power electronics play a crucial role in enhancing the efficiency of e-fuel production. These technologies manage to convert renewable energy sources, such as solar and wind, into electrical energy that can be used for electrolysis and other processes. By precisely controlling the electrical parameters, power electronics ensure that the maximum amount of energy is converted into useful chemical energy, minimizing losses and reducing overall production costs (Pessarrodona et al. 2023).

Moreover, the scalability of e-fuel production is a significant challenge that power electronics can address. Renewable energy sources are inherently variable, and managing this variability is key to scaling up e-fuel production. Power electronics enable the integration of multiple renewable energy sources into a cohesive system, balancing supply and demand in real-time. This capability allows for the expansion of e-fuel production facilities in a modular fashion, making it easier to increase capacity as demand grows without sacrificing efficiency or stability (Barbhuiya and Kanavaris 2024). Storage of e-fuels also presents unique challenges, as the energy content must be preserved over time. Advanced power electronics systems, such as batteries and supercapacitors, are essential for efficient energy storage solutions, which can store excess energy produced during high renewable energy availability periods. These systems ensure that e-fuels can be produced continuously, even when renewable energy generation is intermittent. In addition to these technical challenges, integrating power electronics in e-fuel production supports the development of a more resilient and flexible transport infrastructure. As the transport sector evolves toward greater sustainability, the ability to adapt to different energy sources and storage methods becomes increasingly important. Power electronics facilitate this adaptability by providing the tools to efficiently manage complex energy systems (Kumar and Majid 2020).

Despite the advancements and potential of e-fuel technology, previous research in this field has faced several drawbacks. One major issue has been the high cost and energy intensity of e-fuel production processes, which have limited their commercial viability (Dell'Aversano et al. 2024; Hafez et al. 2023; Wang and Blaabjerg 2020).

Many studies have also highlighted the inefficiencies in the electrolysis process, where significant energy losses occur, making hydrogen production less economical (Erdiwansyah and Husin 2021; Li et al. 2023). Another drawback is the variability and intermittent nature of renewable energy sources, which complicates the continuous production of e-fuels (Saeedmanesh et al. 2018; Failed 2023). While power electronics can help manage this variability, the current technologies have not yet fully optimized this integration, leading to reliability issues (Blaabjerg et al. 2023; El-Shafie et al. 2019). Additionally, the scalability of e-fuel production has been limited by the high initial investment costs and the need for advanced infrastructure (Chan and Satterfield 2020; Luo et al. 2015; Ueckerdt et al. 2021).

Storage solutions for e-fuels also pose challenges. The degradation of stored energy over time and the efficiency of current storage technologies have been significant barriers (Prussi et al. 2021; Agarwal and Valera 2022). Furthermore, the environmental impact of large-scale e-fuel production, particularly concerning the lifecycle emissions of different production pathways, remains a critical area needing more comprehensive assessment (Reddy et al. 2023; Blaabjerg et al. 2015). In summary, while previous research has laid a solid foundation for developing e-fuels, significant challenges remain. High production costs, energy inefficiencies, variability in renewable energy sources, scalability issues, and storage inefficiencies continue to hinder the widespread adoption of e-fuels (Popovic and Ferreira 2010; Kumar et al. 2024). Addressing these challenges through integrating advanced power electronics is crucial for the future success and sustainability of e-fuels as a viable alternative to conventional fossil fuels. This chapter explores these aspects in detail, providing insights into the technologies and strategies that will drive the future of clean transportation (Kumar and Mohanty 2023; Kumar et al. 2020; Dawood 2020).

8.1.1 Background and Motivation

The shift toward sustainable energy solutions in the transport sector is driven by the urgent need to reduce greenhouse gas emissions. Traditional fossil fuels are major contributors to global warming, releasing vast amounts of carbon dioxide and other pollutants into the atmosphere. This environmental impact has accelerated the search for cleaner alternatives, leading to the development of e-fuels. E-fuels, or synthetic fuels, are produced through processes such as electrolysis and using renewable energy sources like solar and wind power. These fuels offer a significant advantage by being compatible with existing internal combustion engines and infrastructure, thus facilitating a smoother transition from fossil fuels to greener options. Furthermore, e-fuels provide a viable pathway to achieving net-zero emissions in the transport sector, addressing climate change and energy security concerns. Integrating advanced power electronics in producing and utilizing e-fuels enhances their efficiency, scalability, and storage, making them a critical component of future sustainable transport solutions.

8.1.2 Scope and Structure of the Chapter

This chapter provides a comprehensive overview of power electronics role in producing e-fuels, covering several key areas for understanding and advancing this technology. It begins with the fundamentals of power electronics in energy conversion, laying the groundwork for how these technologies manage and optimize the transformation of renewable energy into electrical power. The chapter discusses advanced control strategies designed to enhance the performance of power electronics systems, ensuring maximum efficiency and reliability. Following this, integrating renewable energy sources with e-fuel production is explored, highlighting the challenges and solutions in harmonizing variable energy inputs. Scalable production methodologies are examined to address the feasibility of expanding e-fuel production to meet growing demands. Efficient storage solutions for e-fuels are also discussed, focusing on technologies that preserve energy content over time. Finally, the chapter concludes with a look at future directions in power electronics, suggesting pathways for further innovation and development in this critical field.

8.2 Power Electronics in E-Fuel Production

Power electronics underpin modern energy systems, enabling electrical energy conversion, control, and conditioning. In e-fuel production, power electronics play a pivotal role in efficiently integrating renewable energy sources, such as solar and wind, with electrolysis processes to produce hydrogen, a critical e-fuel component. These systems manage renewable energy's variable and intermittent nature, ensuring a stable and continuous electrical power supply necessary for the electrolysis process. By optimizing voltage and current levels, power electronics maximize the efficiency of hydrogen production, minimizing energy losses and reducing overall production costs. Furthermore, advanced power electronic converters and controllers enhance the scalability of e-fuel production facilities, allowing them to adapt to varying energy inputs and demands seamlessly. Through precise control mechanisms, power electronics also facilitate the integration of multiple renewable sources, ensuring a resilient and reliable energy supply chain for sustainable e-fuel production.

Power electronics is crucial in optimizing the electrolysis process for hydrogen production by precisely regulating key electrical parameters such as voltage, current, and frequency. For example, Proton Exchange Membrane (PEM) electrolyzers, commonly used in renewable energy systems, require stable DC voltage and current to operate efficiently and produce high-purity hydrogen. Power electronics systems, including DC/DC converters and active rectifiers, are instrumental in converting and conditioning the power generated from renewable sources, such as solar panels and wind turbines, to meet these stringent requirements. Moreover, power electronics address the inherent variability and intermittency of renewable energy sources. Fluctuations in power supply from solar and wind can adversely affect the performance and reliability of electrolyzers. Power electronics mitigate these fluctuations by employing techniques like real-time voltage regulation and frequency stabilization, ensuring consistent and optimal operating conditions for the electrolysis process. This stability not only enhances hydrogen yield but also significantly improves the overall quality of the hydrogen by minimizing impurities caused by inconsistent electrical inputs. These advancements underscore the transformative potential of power electronics in achieving efficient, reliable, and scalable hydrogen production, particularly when integrated with renewable energy systems.

Modular power electronics systems are essential in addressing scalability challenges in e-fuel production by enabling efficient and cost-effective expansion of production capacity. These systems utilize plug-and-play designs, which allow individual modules to be added or removed seamlessly without disrupting existing operations. For example, additional DC/DC converters or inverters can be integrated in a modular hydrogen production facility as demand grows, enhancing the system's capacity incrementally. This modular approach eliminates the need for extensive infrastructure redesign, significantly reducing costs and complexity during scaling. Producers can adapt to fluctuating market demands or renewable energy inputs by incrementally expanding or modifying their facilities. This flexibility ensures that production remains adaptable and economically viable, particularly in rapidly evolving energy markets. Furthermore, the modular design minimizes downtime during upgrades or maintenance, as individual modules can be serviced or replaced independently. This ensures continuous operation of the production facility, maintaining efficiency and reliability. Combining flexibility, cost-efficiency, and operational resilience, modular power electronics systems provide a robust solution for scaling e-fuel production to meet growing energy demands while maintaining sustainable and reliable operations.

Equation (8.1) highlights the relationship of how precise electrical control, enabled by power electronics, optimizes the production process, ensuring minimal energy loss and maximum yield. The efficiency of the electrolysis process, a cornerstone of e-fuel production, is defined as the ratio of hydrogen energy output to the electrical energy input, expressed as a percentage. Using Faraday's law, the hydrogen production rate is directly proportional to the current supplied, considering the electrolyzer's Faraday efficiency.

The energy conversion efficiency governs the efficiency of the electrolysis process:

$$\eta_{\text{Electrolysis}} = \frac{\text{H}_2 \text{ Energy Output}}{\text{Electrical Energy Input}} \times 100\% \tag{8.1}$$

where: H_2 Energy Output is the energy stored in the produced hydrogen, typically calculated using the lower heating value (LHV) of hydrogen, Electrical Energy Input is the input energy to the electrolyzer.

For a PEM electrolyzer operating at voltage V and current I, the hydrogen production rate is given by Faraday's law in Eq. (8.2):

$$\dot{m}_{H_2} = \frac{\eta_F \cdot n \cdot I}{F} \tag{8.2}$$

where: η_F: Faraday efficiency (dimensionless), n: Number of electrons per reaction (2 for H_2), F: Faraday's constant (96,485 C/mol).

8.2.1 Fundamentals of Power Electronics

Power electronics utilize solid-state devices to efficiently control and convert electrical power, playing a crucial role in e-fuel production systems. The primary components involved are DC/DC converters, AC/DC converters (rectifiers), and DC/AC inverters, each performing distinct functions to ensure optimal performance and integration of renewable energy sources.

DC/DC Converters: DC/DC converters adjust voltage levels within the production process, either stepping up or down as required. In solar-to-hydrogen systems, for instance, these converters regulate the voltage from solar panels to match the specific requirements of the electrolyze. By precisely controlling the voltage, DC/DC converters maximize the efficiency of the electrolysis process, ensuring a consistent and stable hydrogen production rate.

AC/DC Converters: Also known as rectifiers, AC/DC converters transform alternating current (AC) from the grid into direct current (DC), essential for electrolysis. Modern active rectifiers equipped with power factor correction convert AC to DC and enhance overall efficiency by minimizing power losses and reducing harmonic distortion. This results in a cleaner and more efficient power supply, crucial for large-scale industrial applications.

DC/AC Inverters: DC/AC inverters convert direct current back into alternating current for applications such as grid injection or powering AC motors within the production facility. These inverters are vital for integrating renewable energy sources into the grid, allowing excess generated electricity to be fed back into the grid, thereby supporting grid stability and energy sustainability.

Importance in E-Fuel Production: In e-fuel production, the precise control and conversion of electrical power through these devices enable the integration of various renewable energy sources, such as solar and wind, into the production process. By ensuring efficient energy management, these power electronics components contribute to the overall sustainability and scalability of e-fuel production, reducing reliance on fossil fuels and supporting global decarbonization efforts.

By leveraging advanced power electronics, e-fuel production systems can achieve higher efficiency, improved reliability, and greater flexibility, making them a cornerstone of modern renewable energy infrastructure.

8.2.2 Advanced Control Strategies in E-Fuel Production

Advanced control strategies are critical for enhancing the performance of power electronics in e-fuel production systems. These strategies ensure that power converters operate efficiently under varying conditions and loads, which is essential for the reliable and economical production of e-fuels.

Model Predictive Control (MPC): MPC algorithms use real-time models to predict future states of the system and optimize control actions accordingly. By anticipating changes in operating conditions, MPC can dynamically adjust the control inputs to maintain optimal performance. This results in improved dynamic performance, reduced energy losses, and increased efficiency of the e-fuel production process. Model Predictive Control (MPC) optimizes power converter performance by minimizing a cost function that balances system state deviations and control effort over a predictive horizon. By dynamically adjusting control inputs based on future system states, MPC ensures stable and efficient operation under varying conditions, making it critical for integrating renewable energy into e-fuel systems. MPC optimizes the operation of power converters by minimizing a cost function given in Eq. (8.3):

$$J = \sum_{k=0}^{N}\left[Q(x_k - x_{\text{ref}})^2 + R(u_k)^2\right] \tag{8.3}$$

where: x_k: System state at time k, x_{ref}: Reference state, u_k: Control input, Q, R: Weighting matrices, N: Prediction horizon.

The optimization ensures minimal deviation from the reference state while maintaining system stability.

AI-Based Control: Artificial intelligence and machine learning techniques are increasingly integrated into power electronics control systems. These techniques enable predicting energy supply and demand patterns based on historical data and real-time inputs. AI-based control can adaptively manage power converters to optimize their performance, ensuring the e-fuel production system operates at peak efficiency even under fluctuating renewable energy inputs.

Harmonic Suppression Techniques: Harmonic distortion can significantly impact the efficiency and reliability of power electronics systems. Advanced harmonic suppression techniques, such as Active Power Filters (APFs) and Space Vector Modulation (SVM), are employed to minimize these distortions. APFs actively filter out harmonics from the power signal, while SVM optimizes the switching sequences of power converters to reduce harmonic generation. These methods improve system efficiency, enhance power quality, and protect sensitive equipment within the e-fuel production infrastructure.

These advanced control strategies collectively enhance the performance and efficiency of power electronics systems in e-fuel production. Figure 8.1 illustrates the various advanced control strategies, highlighting how they enhance the performance and efficiency of power electronics systems. It illustrates the interconnections among Renewable Energy Sources, Power Electronics, and the Utility Grid/Electrical System, highlighting their mutual interactions through control algorithms. Renewable energy sources such as solar, wind, nuclear, and biofuel generate *uncontrollable power* that may lead to disturbances. Power electronics play a critical role in regulating and converting this power into a more stable form for utility grids. Control algorithms help enhance control strategies by providing disturbance rejection and improving reliability. The utility grid interacts with power electronics to ensure grid stability and manage disturbances such as grid and load variations while maintaining safety and stability under fault conditions and overloads. Feedback pathways represented by blue arrows emphasize reliability improvement and stability, while red pathways indicate disturbances and fault management. Ultimately, the system works as an integrated framework to balance renewable power generation, disturbance rejection, and grid stability. By implementing these strategies, e-fuel production systems can achieve excellent reliability, lower operational costs, and higher energy efficiency, driving the adoption of sustainable fuels in the transport sector.

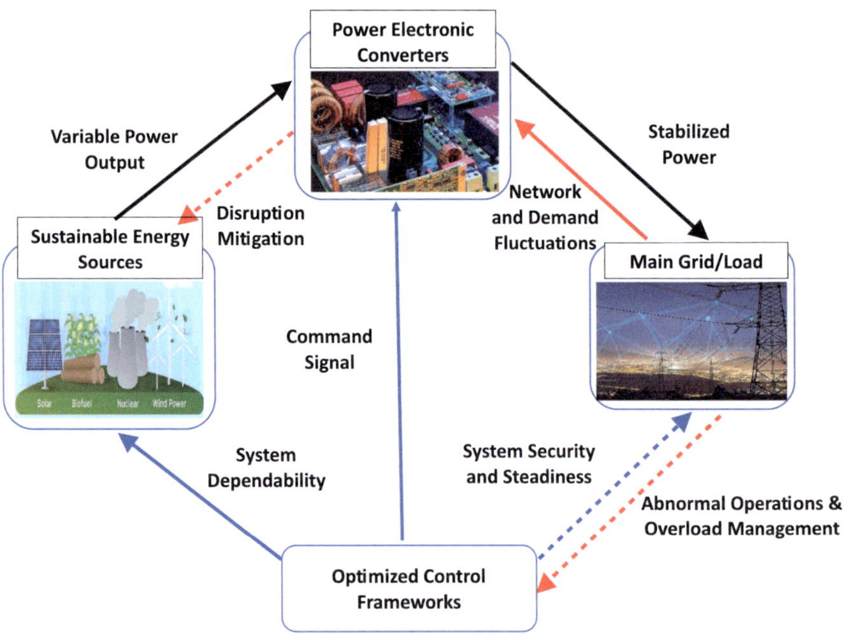

Fig. 8.1 Control algorithms for renewable energy integration and power electronics in utility grids

8.3 Sustainable E-Fuel Production

Achieving sustainability in e-fuel production involves maximizing renewable energy utilization and minimizing waste and emissions.

8.3.1 Renewable Energy Integration

Integrating renewable energy sources, such as solar and wind, with e-fuel production processes is crucial for sustainability. Power electronics solutions ensure the seamless integration of these variable energy sources, providing a stable and continuous energy supply for electrolysis and other production processes.

Maximum Power Point Tracking (MPPT): MPPT algorithms are vital in optimizing the power output from solar panels and wind turbines. These algorithms dynamically adjust the operating point of the power converters to ensure that the renewable energy systems operate at their maximum power point, regardless of changing environmental conditions such as sunlight intensity or wind speed. By continuously tracking and adjusting to the maximum power point, MPPT ensures maximum energy harvest, which is essential for the efficient production of e-fuels. Maximum Power Point Tracking (MPPT) algorithms maximize power output from solar and wind energy systems by dynamically adjusting the operating point to maintain zero power derivative for voltage. This real-time optimization ensures that renewable energy sources operate at peak efficiency, addressing the variability inherent in solar and wind power generation. For photovoltaic (PV) systems, MPPT ensures operation at the maximum power point as per the Eq. (8.4):

$$P = V \cdot I \text{ where} \frac{dp}{dv} = 0 \tag{8.4}$$

Using perturb-and-observe or incremental conductance algorithms, the voltage V and current I are adjusted dynamically to maximize P.

Figure 8.2 describes both the current–voltage (I-V) and power-voltage (PV) curves for a photovoltaic panel under different solar irradiance levels. The simulated photovoltaic (PV) system is rated for a short-circuit current (I_{sc} of 10 A, an open-circuit voltage)V_{oc} of 50 V, and a maximum irradiance of 1000 W/m^2 under standard test conditions. The system parameters include a thermal voltage derived from the diode ideality factor ($n = 1.3$), temperature dependence coefficients ($\alpha I_{sc} = 0.005\,\mathrm{A/^\circ C}$ and $\beta V_{oc} = -0.0025\,\mathrm{V/^\circ C}$), and a temperature of 25 °C. Results show that as solar irradiance increases from 200 to 1000 W/m^2, the short-circuit current (I_{sc}) scales proportionally, while the open-circuit voltage (V_{oc}) remains relatively stable, reflecting the temperature effect on the PV module. The I-V characteristics demonstrate a flat current plateau that sharply drops at V_{oc}, while the PV curves show a distinct peak at the Maximum Power Point (MPP), where power output is maximized. At higher irradiance levels, the current and maximum power output increase,

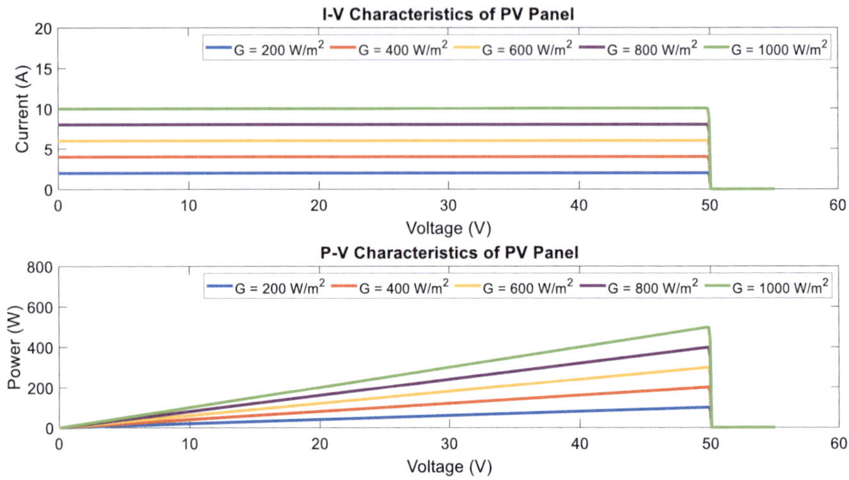

Fig. 8.2 I-V and PV characteristics of a PV panel under varying solar irradiance levels

confirming the PV panel's capability to harness solar energy effectively under variable conditions. This analysis underscores the importance of tracking the MPP to optimize energy generation for practical applications like e-fuel production.

Energy Management Systems (EMS): EMS is critical for coordinating the operation of multiple energy sources, storage systems, and loads within the e-fuel production infrastructure. These systems integrate various renewable energy inputs, manage energy storage devices like batteries and supercapacitors, and control power distribution to the electrolysis and other production processes. EMS ensures stable and efficient operation by balancing the supply from renewable sources with the demand from the e-fuel production process. This real-time balancing act minimizes energy waste and optimizes the overall system performance, making the e-fuel production process more reliable and cost-effective.

Integrating renewable energy sources into e-fuel production processes presents unique challenges due to solar and wind energy's intermittent and variable nature. However, advanced power electronics solutions like MPPT and EMS provide the necessary tools to address these challenges. These systems ensure that renewable energy is efficiently converted and utilized, enhancing the sustainability and viability of e-fuels as an alternative to conventional fossil fuels. Through effective renewable energy integration, power electronics support the production of clean fuels and contribute to the broader goal of reducing greenhouse gas emissions and combating climate change. Figure 8.3 (Chen et al. 1906) shows how solar panels and wind turbines are integrated with the e-fuel production system, with power electronics managing the energy flow and MPPT algorithms optimizing the output. The figure depicts a sustainable energy production and distribution system integrating multiple renewable energy sources, including wind turbines and photovoltaic (PV) arrays, alongside a biogas plant and hydrogen production pathways. Power generated from

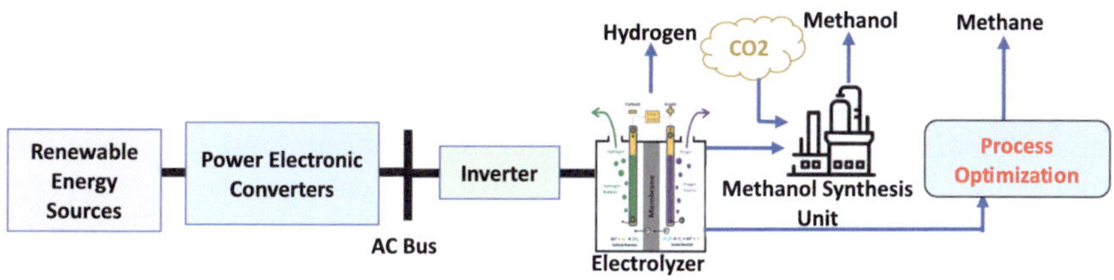

Fig. 8.3 Integrated power-to-methanol and energy pathways for renewable energy utilization (Chen et al. 1906)

wind turbines is converted through an AC/DC/AC converter, while PV arrays use a DC/AC converter to connect to the grid. An electrolyzer utilizes renewable energy to produce hydrogen (blue pathways), which feeds into methanol production, where CO_2 is also utilized. Methanol production and subsequent methodization processes generate methane (yellow pathways) and methanol (red pathways) for industrial use, oil refining, mobility, and electricity generation. The biogas plant provides an additional renewable energy source. This integrated energy framework showcases a closed-loop approach to utilizing renewable energy for multiple applications, ensuring efficient resource utilization, energy conversion, and distribution to industrial, transportation, and power sectors.

8.3.2 Electrolysis for Hydrogen Production

Electrolysis is a key process in hydrogen production, a primary component of many e-fuels. There are several types of electrolyzes, each with its own advantages and challenges, making them suitable for different applications and integration scenarios.

Proton Exchange Membrane (PEM) Electrolyzes: PEM electrolyzes operate at low temperatures and is characterized by high efficiency and rapid response to changes in power supply. This makes them particularly suitable for integration with variable renewable energy sources like solar and wind. PEM electrolyze flexibility and quick response allow them to handle the fluctuations in energy output typical of these sources, ensuring a continuous hydrogen production process. However, they tend to be more expensive due to the materials used in their construction, such as platinum for the catalyst.

PEM Electrolyzers and Power Electronics: Proton Exchange Membrane (PEM) electrolyzers are well-known for responding quickly to variable power inputs, making them suitable for renewable energy systems like solar and wind. However, this rapid response requires highly dynamic power electronics systems. Advanced DC/DC converters are essential to regulate the fluctuating voltage levels from renewable sources, while active rectifiers ensure consistent and stable DC power supply to the electrolyzers. These components work in real-time to accommodate rapid changes in power input, preventing interruptions in hydrogen production and maintaining the high efficiency of the PEM electrolyzer system.

Solid Oxide Electrolyzes: Operating at high temperatures (typically between 700 and 1,000 °C), solid oxide electrolyzes offer high efficiency and can utilize heat from industrial processes. This ability to use waste heat enhances the overall energy efficiency of hydrogen production. Solid oxide electrolyzes can co-electrolyze water and carbon dioxide to produce syngas, a mixture of hydrogen and carbon monoxide, which can be further processed into various e-fuels. The main challenges with this technology are the high operating temperatures, which can lead to material degradation over time, and the need for high thermal energy input.

Solid Oxide Electrolyzers and Power Electronics: Solid Oxide Electrolyzers (SOE) present unique integration challenges due to their high operating temperatures (typically 700–1,000 °C) and the simultaneous need for electrical and thermal energy. Power electronics systems supporting SOE must integrate with thermal management systems to provide both electrical power and thermal stability efficiently. Wide bandgap semiconductors, such as silicon carbide (SiC)-based converters, are particularly well-suited for these applications. These devices can operate at high power densities and elevated temperatures, ensuring efficient energy conversion while maintaining the electrolyzer's demanding thermal requirements.

Alkaline Electrolyzes: Alkaline electrolyzes are the most mature electrolysis technology, known for their robustness and relatively low cost. They use a liquid alkaline electrolyte solution, typically potassium hydroxide, and operate at lower temperatures than solid oxide electrolyzes. Alkaline electrolyzes are well-suited for large-scale hydrogen production due to their durability and lower operational costs. However, they require a more stable power input than PEM electrolyzes, making them less ideal for direct integration with highly variable renewable energy sources.

Alkaline Electrolyzers and Power Electronics: Alkaline electrolyzers operate at lower costs and are more stable than PEM electrolyzers but require steady and reliable power inputs. Power electronics systems for Alkaline electrolyzers are designed for steady-state operation, focusing on maximizing efficiency and minimizing harmonic distortions. These systems include harmonic suppression techniques such as active filters to improve power quality and avoid operational inefficiencies. Because Alkaline electrolyzers are better suited for grid-connected renewable sources with stable power supply, their power electronics systems prioritize reliability over dynamic adaptability, ensuring consistent hydrogen production in such setups.

Each type of electrolyze plays a crucial role in the hydrogen production landscape, contributing to the flexibility and adaptability of e-fuel production systems. The choice of electrolyze technology depends on factors such as the availability of renewable energy sources, cost considerations, and specific application requirements. By leveraging the strengths of different electrolyze types, the hydrogen production process can be optimized to support the sustainable and scalable production of e-fuels, thus significantly reducing carbon emissions in the transport sector. Figure 8.2 illustrates the application of electrolyzes (PEM, Solid Oxide, Alkaline) and how power electronics systems support their operation and integration with renewable energy sources.

For PEM electrolyzers, AI-based control systems have been integrated to dynamically predict and manage variable renewable inputs, ensuring seamless operation under fluctuating power conditions. Harmonic suppression techniques are applied for Alkaline electrolyzers to maintain power quality, reduce energy losses, and ensure operational stability. These examples illustrate how power electronics systems are adapted to the specific requirements of different electrolyzer technologies, ensuring optimal integration and performance.

8.4 Scalable E-Fuel Production

Scalability is essential for adopting e-fuels as a viable alternative to conventional fossil fuels. E-fuels, or synthetic fuels, are produced using renewable energy sources, water, and carbon dioxide, making them a promising solution for reducing greenhouse gas emissions and achieving a carbon–neutral future. To achieve this goal, scalable production processes are crucial to meet the growing demand and to ensure the economic viability of e-fuels. Power electronics play a vital role in enabling the scalability of e-fuel production. These technologies are fundamental in managing and optimizing the flow of electrical energy within the production processes. Power electronics solutions facilitate the efficient conversion and control of electrical power from renewable energy sources, such as solar and wind, to the electrolysis systems used in e-fuel production. This involves using advanced converters, inverters, and control systems to ensure stable and efficient operation. Scaling up e-fuel production from small pilot projects to large industrial-scale operations requires robust and flexible power electronics infrastructure. Modular power electronic systems allow for incremental expansion of production capacity without the need for complete overhauls of existing setups. This modularity enables producers to adapt to varying energy inputs and demands, making the production process more resilient and responsive. Moreover, integrating AI-based control topologies enhances the scalability of e-fuel production. AI algorithms continuously analyze operational data from power electronics systems, such as voltage, current, and load conditions, to make real-time adjustments to power flows. For instance, in a hybrid solar-wind system, AI monitors variations in energy input due to changing environmental conditions and dynamically reallocates power to maximize utilization while avoiding overloading or inefficiencies. During peak solar output, the AI algorithm ensures that excess energy is directed to storage systems or to additional production modules, preventing energy waste. Similarly, during low renewable energy availability, AI prioritizes energy distribution to critical loads, ensuring stable system performance. This smart control ensures that the production process can scale up while maintaining high levels of efficiency and reliability. Scalable e-fuel production hinges on advanced power electronics solutions. These technologies enable the efficient and flexible electrical energy management, facilitating the transition from small-scale pilot projects to large industrial operations. By leveraging modular systems and AI-based controls, the scalability of e-fuel production can be achieved, paving the way for a sustainable and carbon–neutral future in the transport sector. Modular power systems ensure scalability by allowing incremental addition or removal of individual modules. The total power output grows linearly with the number of modules, providing a flexible and efficient approach to scaling e-fuel production facilities as demand increases, without major infrastructure changes. For modular power systems, scalability can be analyzed by summing the contributions of individual modules as the Eq. (8.5) provided below.

$$P_{\text{Total}} = \sum_{i=1}^{N} P_i \tag{8.5}$$

where P_i is the power contribution of the ith module. The modular approach ensures that P_{Total} grows linearly with N, supporting scalability.

Predictive Control with AI in Power Electronics Systems: AI algorithms enable predictive control by identifying patterns in system performance through continuous analysis of historical and real-time operational data. For instance, machine learning models can analyze trends in renewable energy generation, such as solar irradiance or wind speed patterns, alongside energy consumption data. By predicting periods of high energy demand or supply variability, these models allow power electronics systems to proactively adjust settings, such as voltage regulation or energy flow allocation. For example, in a hybrid renewable energy system, AI can predict a drop in solar energy generation due to approaching cloud cover and adjust the power electronics to prioritize stored energy discharge from batteries or supercapacitors. Similarly, during anticipated periods of peak demand, the system can pre-emptively allocate energy to critical loads, ensuring stable operation and minimizing losses. This predictive capability enhances operational efficiency and improves the reliability and resilience of energy systems, making AI an integral component in modern power electronics control.

8.4.1 Modular Power Electronics

Modular power electronics systems facilitate flexible scaling of production facilities. Modules can be seamlessly integrated into the existing system, where each module acts as a self-contained unit that can be connected to other modules. This modular approach enables scalable expansion by adding or removing modules as needed while maintaining overall system stability and operational efficiency. Each module has its own power electronics controls, ensuring it operates independently yet harmoniously within the more extensive system. Modular power electronics are pivotal in achieving scalable e-fuel production, which is essential for meeting the growing demand for sustainable energy sources. E-fuels, synthesized from renewable energy, water, and carbon dioxide, promise a carbon–neutral future. Still, their production must scale efficiently from pilot projects to industrial-scale operations to be economically viable.

Modular power electronics provide the flexibility and efficiency needed for this scalability. These systems consist of interconnected power electronic modules that can be easily expanded or reconfigured to match production requirements. This modularity allows for a gradual increase in production capacity, accommodating varying energy inputs from renewable sources like solar and wind without extensive redesigns or overhauls of the existing infrastructure. Modular systems in real-world applications are provided below:

1. *Plug-and-Play Solar-to-Hydrogen Facilities*: Modular power electronics are used in solar-to-hydrogen facilities, where each module consists of a DC/DC converter, MPPT controller, and PEM electrolyzer. These modules can be added or removed based on the required production capacity.
2. *Distributed Microgrid Systems*: Modular power electronics systems in microgrids include individual energy storage and distribution units, such as battery packs or supercapacitors. Each module operates independently and communicates with other modules through a centralized Energy Management System (EMS).
3. *Hydrogen Production with Modular Electrolyzers*: Facilities using modular electrolyzers integrate multiple self-contained units, each equipped with its own power electronics. These units collectively scale hydrogen production by adding modules as needed based on energy availability and production demand.

The key advantage of modular power electronics lies in their adaptability. As production needs grow, additional modules can be integrated seamlessly, enhancing the overall system's capacity and reliability. This ensures a continuous and stable electrical power supply to the electrolysis systems, crucial for e-fuel production. Modular systems also simplify maintenance and upgrades, as individual modules can be serviced or replaced without disrupting the entire operation. Furthermore, integrating AI-based control topologies with modular power electronics optimizes their performance. AI algorithms can dynamically adjust the operation of these systems, improving energy efficiency and reducing losses. This intelligent control ensures that the production process remains efficient and reliable even as it scales up. In summary, modular power electronics are fundamental to scalable e-fuel production. They offer the flexibility to expand production capacity incrementally, ensuring efficient and reliable operation. Coupled with AI-based controls, these systems enable the sustainable and economical production of e-fuels, supporting the transition to a carbon–neutral energy landscape.

Plug-and-Play Modules: Standardized modules that can be easily integrated or removed, allowing rapid scaling and maintenance. Plug-and-play modules in power electronics are revolutionizing scalable e-fuel production, which is crucial for transitioning to sustainable energy. E-fuels, synthesized using renewable energy, water, and carbon dioxide, need scalable production systems to meet increasing demand and ensure economic viability. Plug-and-play modules offer a flexible, efficient solution for scaling e-fuel production. These pre-configured, self-contained units can be easily added to or removed from the system, allowing for seamless capacity expansion without significant infrastructure changes. This modular approach enables producers to start with small-scale operations and incrementally increase production as needed.

The key advantage of plug-and-play modules is their ease of integration. As production demands grow, additional modules can be integrated effortlessly, enhancing system capacity and reliability. This adaptability ensures a stable supply of electrical power to electrolysis systems, which is vital for e-fuel synthesis. Moreover, modularity simplifies maintenance and upgrades, as individual units can be serviced or replaced without halting overall operations. Integrating AI-based control topologies with these modules further optimizes performance. AI algorithms can dynamically adjust module operations, improving efficiency and reducing energy losses. This intelligent control maintains efficiency and reliability even as production scales up.

Redundancy and Reliability: Modular systems improve reliability by allowing redundancy, where failed modules can be bypassed without affecting overall operation. Redundancy and reliability in power electronics are critical for scalable e-fuel production, ensuring consistent and efficient operation as production scales from pilot projects to industrial levels. E-fuels, produced using renewable energy sources, water, and carbon dioxide, require robust systems to meet growing demand sustainably. Redundancy in power electronics involves incorporating multiple backup components within the system. In a modular power electronics system, each module has independent control mechanisms designed to handle a specific portion of the total load. In a module failure, the system's intelligent control architecture detects the fault and redistributes the load among the remaining operational modules. This process ensures uninterrupted operation and maintains system stability. For example, in a modular hydrogen production facility, if one DC/DC converter module fails, the power management system dynamically adjusts the output of the remaining converters to compensate for the lost capacity without disrupting the hydrogen production process. In scalable e-fuel production, redundancy is vital for maintaining continuous operation, significantly as the system expands. It minimizes downtime and prevents production losses, ensuring a stable supply of e-fuels. Reliability is equally essential, guaranteeing the system's performance under various conditions. High-quality power electronic components and advanced control systems enhance reliability, providing consistent energy conversion and management. Reliable systems are essential for handling the variable nature of renewable energy sources like solar and wind, ensuring efficient and stable power supply to electrolysis units critical for e-fuel synthesis. Integrating AI-based control systems with redundant power electronics further boosts reliability. AI algorithms can predict potential failures and dynamically adjust operations to mitigate risks, enhancing overall system resilience.

8.4.2 Grid Integration

Seamless integration with the electrical grid is critical for scalable e-fuel production, ensuring efficient and sustainable operations. E-fuels, synthesized using renewable energy sources, water, and carbon dioxide, require flexible and resilient systems to manage the bidirectional power flow between production facilities and the grid.

Advanced power electronics solutions are essential in achieving this seamless integration.

Grid-forming inverters play a pivotal role in this integration. These inverters can operate in island and grid-synchronized modes, providing significant flexibility and resilience. In island mode, grid-forming inverters can independently manage the power supply, ensuring continuous operation of e-fuel production even during grid outages. This capability is particularly important for maintaining production stability and preventing downtime. When synchronized with the grid, these inverters ensure smooth power exchange, maintaining system stability and optimizing energy flow.

Demand response management is another crucial grid integration aspect for e-fuel production. Advanced power electronics solutions enable real-time adjustments in production processes based on grid demand.

Scenario of Demand Response Management in E-Fuel Production: During periods of high electricity demand on the grid, such as peak hours in the evening when residential consumption is at its highest, the e-fuel production facility can participate in demand response programs by temporarily reducing its energy consumption. This is achieved by scaling down non-critical operations or adjusting the production schedule of e-fuels. The energy the facility would have used is thus made available to the grid, helping alleviate strain and prevent potential overloads. Conversely, during low grid demand or when there is excess generation from renewable energy sources, like a sunny, windy afternoon, the facility can increase its e-fuel production. Advanced power electronics systems enable the facility to absorb the surplus energy, which might otherwise be curtailed, and utilize it for synthesizing e-fuels. This optimizes energy usage within the facility and contributes to overall grid stability by balancing supply and demand. Additionally, energy storage systems integrated with the facility can be charged during periods of excess renewable energy and discharged when needed. For instance, stored energy can maintain production levels during peak grid demand without drawing additional power from the grid.

For instance, during periods of high grid demand, production can be scaled down, while excess renewable energy can be utilized for e-fuel synthesis during low demand. This approach balances the load on the grid and maximizes the use of renewable energy, aligning with sustainability goals. Moreover, bidirectional power flow facilitated by advanced power electronics allows production facilities to contribute to grid stability. During periods of excess production, surplus power can be fed back into the grid, supporting overall grid resilience. Conversely, facilities can draw power from the grid during production shortfalls, ensuring continuous operation.

The seamless integration with the electrical grid is essential for scalable e-fuel production. Advanced power electronics solutions enable flexible and resilient operations, including grid-forming inverters and demand response management. These technologies ensure efficient bidirectional power flow, optimizing energy usage, reducing costs, and supporting the stability and sustainability of both the production process and the electrical grid.

Figure 8.4 illustrates how e-fuel production facilities interact with the electrical grid, showing the role of grid-forming inverters and demand response management. The figure illustrates a distributed energy management system integrating various

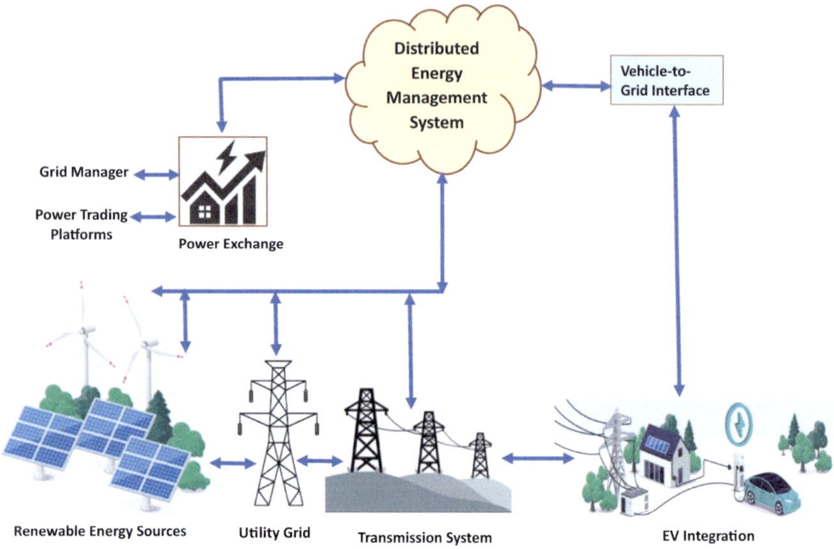

Fig. 8.4 Distributed energy management system framework for energy management and trading

resources to optimize energy management and grid stability. Renewable energy sources such as solar panels and wind turbines contribute power to the distributed energy management system (DEMS), which coordinates with the grid infrastructure and network operators. The DEMS also facilitates energy trading within energy markets, enabling financial transactions and market participation. A key component is the Vehicle-to-Grid (V2G) Gateway, which allows electric vehicles (EVs) to act as flexible storage systems, supplying power back to the grid when needed. The DEMS centrally manages the energy flow between all connected entities, including renewable sources, grid infrastructure, and EVs, ensuring efficient energy utilization, grid stability, and market integration. This interconnected ecosystem highlights the role of advanced digital platforms in modern energy systems, promoting sustainability and flexibility.

8.4.3 Distributed Generation and Microgrids

Distributed generation and microgrids provide significant flexibility and resilience to e-fuel production systems, which is essential for maintaining efficiency and reliability. By decentralizing energy production, these systems can minimize transmission losses and enhance the reliability of the energy supply, making them ideal for supporting e-fuel production facilities. In microgrids, power balance equations ensure that generation and storage meet consumption and losses, enabling a stable energy supply. The state of charge (SoC) of batteries dynamically updates based on charging and discharging activities, facilitated by power electronics to optimize energy usage and maintain system reliability. In a microgrid system, the power balance equation can be presented as provided in Eq. (8.6):

$$P_{\text{gen}} + P_{\text{storage}} = P_{\text{Load}} + P_{\text{loss}} \tag{8.6}$$

where: P_{gen}: Power generated by renewable sources, P_{storage}: Net power from energy storage systems, P_{Load}: Power consumed, P_{loss}: Transmission and conversion losses.

Whereas the state of charge (SoC) of the battery is updated using Eq. (8.7):

$$SoC(t) = SoC(t-1) + \frac{P_{\text{charge}} - P_{\text{discharge}}}{C_{\text{battery}}} \tag{8.7}$$

where: C_{battery}: Battery capacity.

Microgrid controllers are central to the operation of microgrids. These sophisticated systems manage multiple distributed energy resources (DERs), such as solar panels, wind turbines, and small-scale hydroelectric plants. Microgrid controllers ensure that these diverse energy sources operate harmoniously, maintaining a stable and efficient energy supply. They balance supply and demand within the microgrid, optimize energy usage, and seamlessly handle transitions between grid-connected and island modes. This capability is crucial for e-fuel production facilities, which require a consistent and reliable energy supply to maintain continuous operations.

Energy storage integration within microgrids further enhances their ability to manage the variability of renewable energy sources. Energy storage systems, such as batteries and flywheels, can store excess energy generated during periods of high renewable output and release it when production is low or demand is high. This integration ensures that the microgrid remains stable and can provide a reliable energy supply, regardless of fluctuations in renewable energy generation. For e-fuel production, this means uninterrupted operations and optimized energy use, reducing reliance on the main grid and enhancing overall efficiency. The accompanying figure illustrates a typical microgrid setup with distributed generation. It highlights the critical role of microgrid controllers in coordinating the operation of various DERs and integrating energy storage systems.

The distributed generation and microgrids offer crucial advantages for e-fuel production systems. Microgrid controllers and energy storage integration are key to managing variable renewable energy sources, reducing transmission losses, and enhancing energy reliability. These systems ensure that e-fuel production facilities can operate efficiently and continuously, supporting a sustainable and resilient energy future. Figure 8.4 showcases how these components work together to create a resilient and efficient energy network that supports e-fuel production.

8.5 Storable E-Fuel Solutions

Efficient and safe storage of e-fuels is a significant challenge that must be addressed to realize their potential as a sustainable energy source. E-fuels, synthesized from renewable energy, water, and carbon dioxide, require advanced storage solutions to remain stable, accessible, and safe. Power electronics play a critical role in developing these effective storage systems. Power electronics solutions are essential for managing the conversion and conditioning of energy necessary for e-fuel production and storage. They ensure that the energy used to synthesize e-fuels is efficiently managed, minimizing losses and optimizing production. This is particularly important for storing e-fuels, as inefficiency can increase costs and reduce overall effectiveness. Efficient hydrogen storage requires precise control of compression and liquefaction processes. The compression energy depends on the logarithmic ratio of final to initial pressure, while liquefaction requires accounting for the specific heat capacity over the temperature range. Advanced power electronics regulate these processes, minimizing energy consumption and ensuring safe, stable hydrogen storage. For hydrogen compression, the energy required can be given as per Eq. (8.8):

$$E_{\text{compression}} = \frac{\gamma}{\gamma - 1} \cdot P \cdot V \cdot \ln\left(\frac{P_f}{P_i}\right) \tag{8.8}$$

where: P_f, P_i: Final and initial pressures, V: Volume of hydrogen, γ: Heat capacity ratio.

Equation (8.9) provides the energy requirement for hydrogen liquefaction using the specific enthalpy:

$$E_{\text{Liquefaction}} = \int_{T_i}^{T_f} C_p dT \tag{8.9}$$

where: C_p: Specific heat capacity, T_i, T_f: Initial and final temperatures.

Advanced power electronics systems enable precise control over the storage conditions of e-fuels. This includes maintaining optimal temperatures, pressures, and chemical conditions to ensure the stability and safety of stored e-fuels. For example, power electronic controllers can regulate the operation of compressors and cooling systems in hydrogen storage, ensuring that hydrogen remains in its desired state without risking leaks or degradation.

Energy management systems (EMS), powered by advanced power electronics, can optimize the use of stored e-fuels. These systems can dynamically manage the release of e-fuels based on demand, ensuring that the fuel is available when needed and minimizing waste. This real-time control and monitoring capability enhances e-fuel storage and usage efficiency.

Safety is another critical aspect to which power electronics contribute significantly. Advanced sensors and control systems can detect anomalies in storage conditions, such as pressure drops or temperature spikes, and take immediate corrective actions to prevent accidents. This proactive approach ensures the safe handling and storage of e-fuels, protecting the infrastructure and the environment.

The efficient and safe storage of e-fuels is a complex challenge that benefits significantly from advanced power electronics solutions. These technologies provide precise control, optimize energy management, and enhance safety, ensuring that e-fuels can be stored effectively and used efficiently. By addressing these storage challenges, power electronics pave the way for the widespread adoption of e-fuels, supporting a sustainable and resilient energy future.

Optimization of Charging and Discharging Cycles: Power electronics systems, such as batteries and supercapacitors, are critical in optimizing an energy storage system's charging and discharging cycles. These systems utilize advanced control algorithms to regulate current and voltage during energy transfer processes. By ensuring that energy storage devices operate within their optimal parameters, power electronics help maximize energy utilization while minimizing losses. This improves overall efficiency and extends the operational lifespan of energy storage components. For instance, during the charging process, precise control prevents overcharging, while during discharging, it ensures consistent energy delivery without exceeding safe operating limits.

Role of Energy Management Systems (EMS): Integrated with advanced power electronics, Energy Management Systems (EMS) coordinate energy flow between storage and production units to ensure seamless operations. EMS dynamically allocates energy in real-time based on supply and demand conditions. For example, during surplus renewable energy generation periods, EMS prioritizes charging the energy storage systems while concurrently supplying power to production units such as electrolyzers. Conversely, during low renewable energy availability periods, EMS manages the controlled discharge of stored energy to maintain continuous production. This dynamic management stabilizes the energy supply, enhances system efficiency, and reduces dependency on external energy sources or backup systems.

Practical Examples of EMS in E-Fuel Production Systems: EMS coordinates the energy transfer between solar panels, storage systems, and electrolyzers in a solar-to-hydrogen facility. During peak solar power generation, EMS directs surplus energy to storage devices, ensuring electrolyzers operate optimally. When solar generation dips, the stored energy is intelligently discharged to maintain uninterrupted hydrogen production, this coordination effectively balances the inherent fluctuations of solar energy, demonstrating the practical application of EMS in enhancing the efficiency and reliability of integrated e-fuel production systems.

8.5.1 Hydrogen Storage

Hydrogen, a primary component of e-fuels, requires specialized storage solutions due to its low density and high flammability. Efficient and safe hydrogen storage is essential for its role in a sustainable energy future. Power electronics systems are integral in managing hydrogen compression, liquefaction, and storage, ensuring safety and efficiency. The chapter explains how power electronics systems manage hydrogen compression and liquefaction processes, ensuring optimal safety and efficiency. Advanced power electronic controllers regulate compressors and cryogenic systems, reducing energy consumption and operational costs while maintaining the integrity of hydrogen storage systems. Monitoring systems powered by power electronics detect anomalies in pressure and temperature, enabling real-time corrective actions to prevent accidents. As a practical example, a solar-to-hydrogen production plant using Proton Exchange Membrane (PEM) electrolyzers provides a compelling example of how power electronics enhance hydrogen storage systems. In such facilities, power electronics are critical in controlling high-efficiency compressors that pressurize hydrogen for storage, ensuring optimal energy use and operational precision. Additionally, advanced monitoring systems powered by power electronics continuously detect and address potential issues, such as pressure drops or temperature fluctuations, thereby safeguarding the storage process. These integrated systems not only improve energy efficiency but also enhance the overall reliability of operations, enabling the safe and sustainable storage of hydrogen for various applications, including transportation and energy production.

(a) Electrolyzes are the initial step in hydrogen production, converting water into hydrogen and oxygen using electrical energy. Two main types of electrolyzes are prominent in hydrogen production:

PEM Electrolyzers (Proton Exchange Membrane): These electrolyzes offer high efficiency and rapid response times, making them suitable for integration with dynamic renewable energy sources such as solar and wind. Their ability to quickly adjust to fluctuating power inputs ensures efficient hydrogen production, aligning supply with the variable nature of renewable energy.

Solid Oxide Electrolyzers: Operating at high temperatures, these electrolyzers are highly efficient and ideal for industrial-scale hydrogen production. They can utilize excess heat from industrial processes, improving overall energy efficiency.

(b) Compression and liquefaction are critical for reducing the volume of hydrogen, making it easier to store and transport.

High-Efficiency Compressors: Advanced power electronics control the hydrogen compression, significantly reducing energy consumption and enhancing safety. These systems ensure that hydrogen is compressed efficiently, minimizing the energy required and reducing operational costs.

Scenario of Demand Response Management in E-Fuel Production: During periods of high electricity demand on the grid, such as peak hours in the evening when residential consumption is at its highest, the e-fuel production facility can participate in demand response programs by temporarily reducing its energy consumption. This is achieved by scaling down non-critical operations or adjusting the production schedule of e-fuels. The energy that the facility would have used is thus made available to the grid, helping to alleviate strain and prevent potential overloads. Conversely, during low grid demand or when there is excess generation from renewable energy sources, like a sunny, windy afternoon, the facility can increase its e-fuel production. Advanced power electronics systems enable the facility to absorb the surplus energy, which might otherwise be curtailed, and utilize it for synthesizing e-fuels. This optimizes energy usage within the facility and contributes to overall grid stability by balancing supply and demand. Additionally, energy storage systems integrated with the facility can be charged during periods of excess renewable energy and discharged when needed. For instance, stored energy can maintain production levels during peak grid demand without drawing additional power from the grid.

Cryogenic Systems: Power electronics also manage the liquefaction of hydrogen, which involves cooling it to extremely low temperatures (−253 °C or −423°F) to convert it into a liquid form. Cryogenic systems, equipped with advanced controls, ensure this process's stable and efficient operation, allowing for the compact storage of liquid hydrogen.

(c) Safety and efficiency are paramount in hydrogen storage. Power electronics systems provide precise control over the compression and liquefaction processes, ensuring that hydrogen is stored optimally. Advanced monitoring and control systems can detect anomalies, such as pressure fluctuations or temperature deviations, and take immediate corrective actions to prevent accidents.

Hydrogen storage is a complex challenge that benefits significantly from advanced power electronics solutions. Power electronics ensure efficient and safe hydrogen production and storage by managing electrolyzes, compression, and liquefaction processes. These technologies are critical for integrating hydrogen into the energy grid, supporting the broader adoption of e-fuels, and contributing to a sustainable energy future. Figure 8.5 illustrates the hydrogen storage process, including compression, liquefaction, and the role of power electronics in managing these systems. The figure represents a hydrogen energy cycle that starts with energy production from renewable sources like solar energy. This energy is used in hydrogen production via an electrolyzer, where water is split into hydrogen (H_2) and oxygen. The generated hydrogen is then moved to hydrogen storage, ensuring its availability for future use. In the next step, hydrogen is consumed through a fuel cell, which converts chemical energy into electrical energy. Finally, this electricity is utilized to power applications such as electric vehicles, highlighting the role of hydrogen as a clean and sustainable energy carrier. The flowchart emphasizes the integration of renewable energy, hydrogen storage, and fuel cell technology to enable clean energy solutions for transportation and other end-uses.

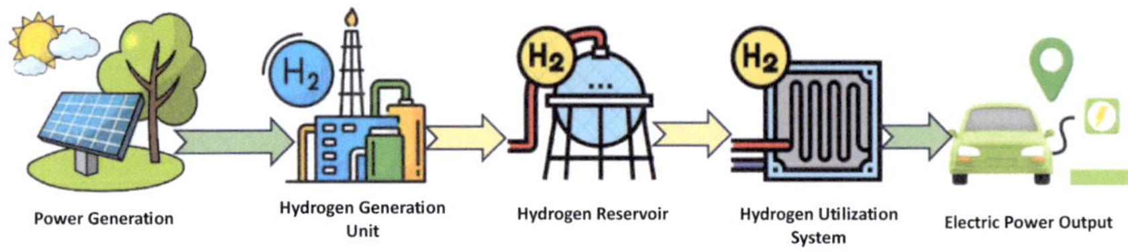

Fig. 8.5 Hydrogen energy cycle: from production to electricity generation

8.5.2 *Synthetic Fuel Storage*

Synthetic fuels, produced from hydrogen and captured carbon dioxide, offer a sustainable alternative to conventional fossil fuels. Efficient storage solutions are essential to maintain their stability and energy density, ensuring these fuels remain viable for various applications. Power electronics systems play a critical role in controlling the storage conditions for synthetic fuels, ensuring optimal performance and safety. The chapter highlights how power electronics manage temperature and pressure controls while storing synthetic fuels to maintain stability and energy density. It describes the role of power electronics in monitoring containment integrity, using advanced sensors to detect leaks or structural issues, thus enhancing safety. The integration of power electronics ensures energy-efficient storage by optimizing cooling and heating mechanisms to prevent the degradation of synthetic fuels. As a practical example, A Fischer–Tropsch synthesis plant producing synthetic diesel demonstrates the critical role of power electronics in synthetic fuel production and storage. During synthesis, power electronics regulate reaction conditions, such as temperature and pressure, to optimize efficiency and ensure consistent output quality. Additionally, storage tanks equipped with electronically controlled temperature and pressure systems powered by advanced electronics maintain the stability and energy density of the synthetic fuel over extended periods. These systems significantly enhance the overall efficiency of the production and storage processes, reducing degradation and minimizing energy losses, making synthetic fuels a more viable and sustainable alternative to conventional fossil fuels. In Fischer–Tropsch synthesis, the stoichiometric ratio of hydrogen to carbon monoxide determines the hydrocarbon output, with efficiency measured as the ratio of energy in the hydrocarbons to the input energy. Similarly, methanation converts hydrogen and carbon dioxide into methane, with power electronics optimizing reaction conditions to maximize efficiency and minimize energy losses. In Fischer–Tropsch reaction the stoichiometry of the reaction can be presented as in Eq. (8.10):

$$(2n + 1)H_2 + nCO \rightarrow C_nH_{2n+2} + H_2O \tag{8.10}$$

The efficiency of the synthesis is given by Eq. (8.11):

$$\eta_{\text{FT}} = \frac{\text{Energy in hydrocarbons}}{\text{Energy in Reactants}} \tag{8.11}$$

In case of the methanation hydrogen reaction takes place as per Eq. (8.12).

$$CO_2 + 4H_2 \rightarrow CH_4 + 2H_2O \tag{8.12}$$

Here the energy efficiency can be expressed as in Eq. (8.13).

$$\eta_{\text{methanation}} = \frac{\text{Energy Contant of } CH_4}{\text{Input Energy for Reaction}} \tag{8.13}$$

(a) Power-to-liquid (PtL): Processes are central to the production of synthetic fuels. These processes convert hydrogen and carbon-based feedstocks into liquid fuels, suitable for use in existing infrastructure.

Fischer–Tropsch Synthesis: This process converts hydrogen and carbon monoxide into liquid hydrocarbons like synthetic diesel and kerosene. Power electronics systems are crucial in controlling the reaction conditions temperature, pressure, and catalyst activity to optimize yield and efficiency. By precisely managing these variables, power electronics ensure consistent and high-quality production of synthetic fuels.

Methanation: This process produces synthetic methane by reacting hydrogen with carbon dioxide. The efficiency of methanation depends heavily on maintaining precise control over temperature and pressure. Power electronics enable real-time adjustments to these parameters, ensuring maximum efficiency and optimal production rates. This precision is essential to minimize energy consumption and enhance the overall sustainability of synthetic fuel production. Figure 8.5 shows a schematic of a solar-to-hydrogen production plant, including solar panels, MPPT controllers, and PEM electrolyzers.

(b) Storage Solutions for synthetic fuels require careful management to maintain their stability and energy density. Power electronics systems control storage conditions, such as temperature, pressure, and containment integrity.

Temperature and Pressure Control: Synthetic fuels must be stored at specific temperatures and pressures to prevent degradation and maintain their energy content. Power electronics systems regulate cooling and heating mechanisms, ensuring that storage conditions remain within optimal ranges. This control is vital to preserving the quality and usability of synthetic fuels over extended periods.

Containment Integrity: Advanced sensors and control systems, powered by electronics, monitor the structural integrity of storage tanks and vessels. These systems detect any leaks or weaknesses and initiate corrective actions to prevent accidents, ensuring safe storage.

(c) Safety and efficiency are paramount in synthetic fuel storage. Power electronics systems enhance safety by providing precise control and monitoring capabilities, reducing the risk of accidents. Additionally, these systems improve efficiency by optimizing energy usage in storage processes, minimizing losses, and maximizing the economic viability of synthetic fuels. The importance of precise control provided by power electronics systems in reducing energy losses and maximizing operational efficiency is emphasized. The use of real-time energy management systems powered by power electronics is detailed, showcasing their ability to dynamically adjust storage conditions for optimal performance. For a practical example, an offshore wind-powered hydrogen production and storage plant exemplifies the practical application of power electronics in sustainable energy systems. In this facility, power electronics efficiently manage the energy flow from wind turbines to electrolyzers, ensuring the optimal wind energy conversion into hydrogen. Advanced electronic controls further enhance the liquefaction and storage processes by maintaining precise operational parameters, enabling the safe handling of hydrogen even in challenging offshore environmental conditions. These technological advancements result in high energy conversion efficiency and a robust storage infrastructure, demonstrating the viability of large-scale, sustainable hydrogen production for diverse applications.

Synthetic fuel storage is a complex and critical aspect of sustainable energy solutions. Power electronics systems ensure stored synthetic fuels' stability and energy density by controlling temperature, pressure, and containment integrity. Through precise management of PtL processes and storage conditions, power electronics contribute significantly to the efficiency, safety, and sustainability of synthetic fuel production and storage. Figure 8.6 depicts synthetic fuel storage solutions, showing synthesis and methanation processes and how power electronics ensure optimal storage conditions. The figure outlines the synthetic fuel production process from renewable energy sources to end-use applications. It begins with electricity generation using renewable sources like wind, solar, and hydropower. This electricity powers the hydrogen production process through electrolysis, where water (H_2O) is split into hydrogen (H_2) and oxygen. The next step involves CO_2 supply, sourced from the atmosphere, biomass, or industrial emissions. Hydrogen and CO_2 are combined in a chemical synthesis to produce synthetic fuels. These fuels undergo refining for further processing into e-fuels. The final synthetic fuels are distributed for various applications, including diesel fuel for motor vehicles and marine transport, jet fuel for aviation, methane for energy supply, and heating and chemical production. This integrated approach highlights a sustainable pathway to produce clean fuels while reducing carbon emissions.

Offshore Wind-to-Fuel Initiative: The offshore wind-to-fuel initiative showcases the use of advanced power electronics systems to harness the consistent wind resources available in offshore locations for large-scale hydrogen production. Wind turbines convert wind energy into electricity, which is then efficiently managed by power electronics to ensure stable and uninterrupted hydrogen production. These

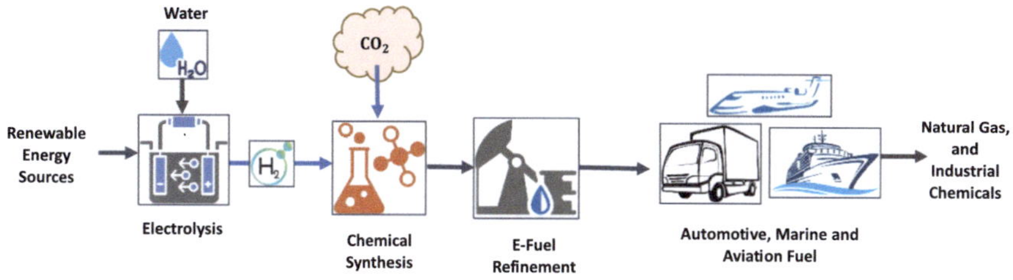

Fig. 8.6 Pathway for synthetic fuel production from renewable energy

systems regulate the variability of wind energy, a challenge inherent to renewable sources, by dynamically controlling the flow of electricity to Proton Exchange Membrane (PEM) electrolyzers or Solid Oxide Electrolyzers (SOE), depending on the system design. This ensures optimal electrolyzer performance even under fluctuating energy conditions. The initiative offers significant benefits, particularly the abundance and reliability of offshore wind resources compared to onshore locations. This consistency enables large-scale hydrogen production, making it a viable option for meeting growing energy demands. Moreover, offshore facilities often avoid land-use conflicts and benefit from higher wind speeds, further enhancing energy conversion efficiency.

However, the project faces several challenges. One major obstacle is the high cost of offshore infrastructure, including installing and maintaining wind turbines and hydrogen production facilities in harsh marine environments. Additionally, transporting hydrogen to onshore locations poses logistical complexities, particularly in ensuring its safe and efficient handling. Advanced liquefaction systems, powered by precise electronic controls, are crucial for converting hydrogen into a compact liquid form, facilitating easier storage and transportation.

Using grid-forming inverters is another critical element of this initiative, enabling the seamless integration of the offshore system with the grid. These inverters provide the necessary flexibility for bidirectional energy flows and ensure stable grid operations, even during high wind variability.

Application of Power Electronics: Power electronics systems manage the variability of wind energy and ensure stable hydrogen production by integrating the fluctuating energy into electrolyzer systems. Wide bandgap semiconductors (e.., silicon carbide-based converters) enhance the performance of these systems.

Achievements:

- Improved energy conversion efficiency.
- Compact, high-performance designs enabled by wide bandgap semiconductors.
- Reduced downtime through advanced control systems.

8.6 Other Power Projects

This section presents studies of integrated power electronics solutions in e-fuel production, their feasibility, and their benefits.

8.6.1 *Solar-to-Hydrogen Projects*

Solar-to-hydrogen projects represent a promising pathway toward sustainable and scalable e-fuel production, leveraging the abundant energy from the sun to produce hydrogen, a clean and versatile fuel. These projects integrate solar power with hydrogen production using advanced power electronics systems, optimizing efficiency and demonstrating the potential for large-scale implementation. The core concept involves utilizing photovoltaic (PV) panels to convert sunlight into electrical energy. High-efficiency PV panels equipped with Maximum Power Point Tracking (MPPT) controllers are employed to maximize the capture and conversion of solar energy. MPPT controllers are crucial as they continuously adjust the electrical load to ensure the PV panels operate at their optimal power point, thus enhancing overall efficiency.

The electricity generated by the PV panels is then used to power Proton Exchange Membrane (PEM) electrolyzers. These electrolyzers split water into hydrogen and oxygen through an electrochemical process. PEM electrolyzers are preferred for their high efficiency and rapid response, making them suitable for dynamic renewable energy inputs. Advanced power electronics systems play a vital role in this setup, controlling the operation of the MPPT controllers and the electrolyzers to ensure stable and efficient hydrogen production.

8.6.2 *Desert Solar Hydrogen Production Plant*

This plant, located in a high-sunlight desert region, is a practical example of solar-to-hydrogen integration. The system uses high-efficiency PV panels with MPPT controllers to harness solar energy effectively. The electricity produced powers PEM electrolyzers, controlled by advanced power electronics, to generate hydrogen. The plant has achieved high energy conversion efficiency, demonstrating the feasibility of large-scale hydrogen production from solar energy. The results underscore the potential of solar-to-hydrogen projects to contribute significantly to the global renewable energy landscape. The schematic of a solar-to-hydrogen production plant includes key components, such as solar panels, MPPT controllers, and PEM electrolyzers, integrated into a cohesive system. This setup ensures the efficient conversion of solar energy into hydrogen, ready for storage or use as a clean fuel.

The Desert Solar Hydrogen Plant demonstrates the integration of renewable energy systems with hydrogen production, leveraging the abundant sunlight available in desert regions. This facility utilizes high-efficiency photovoltaic (PV) panels equipped with Maximum Power Point Tracking (MPPT) controllers to maximize energy capture. MPPT controllers dynamically adjust the electrical load to ensure that the PV panels operate at their optimal power point, thus enhancing energy conversion efficiency. The electricity generated is then fed into Proton Exchange Membrane (PEM) electrolyzers, which split water into hydrogen and oxygen with high efficiency and responsiveness. PEM electrolyzers are particularly well-suited for this application due to their rapid adaptability to variable energy inputs and their ability to deliver high-purity hydrogen. The plant's operational advantages include the efficient use of intense and consistent sunlight in desert regions, leading to a high energy yield and cost-effective hydrogen production. Additionally, the modular nature of the system enables incremental scaling as energy demands grow, making it a flexible solution for expanding hydrogen production.

However, the plant also faces significant challenges. Managing heat dissipation in extreme desert climates is critical to maintaining the performance and longevity of both the PV panels and the electronic components. Overheating can reduce the efficiency and lifespan of these systems. Ensuring the reliability of components under harsh environmental conditions, such as dust storms and temperature fluctuations, is another major challenge. The plant's design incorporates robust cooling systems and protective measures to address these issues, but they add to the complexity and operational costs.

Application of Power Electronics: Advanced power electronics, including Maximum Power Point Tracking (MPPT) controllers and high-efficiency DC/DC converters, are used to maximize energy capture from solar panels and regulate the input to Proton Exchange Membrane (PEM) electrolyzers.

Achievements:

- Significant energy savings by minimizing energy conversion losses.
- Modular design enables incremental scalability, allowing the facility to expand production capacity.

8.6.3 Wind-to-Fuel Initiatives

Wind-to-fuel initiatives are emerging as a promising solution for sustainable energy production, converting wind energy into e-fuels like hydrogen or synthetic hydrocarbons. These projects utilize the abundant and renewable energy generated by wind turbines to produce clean fuels, which can be stored, transported, and used in various applications, thereby supporting the global transition to a low-carbon economy.

Core Concept: The fundamental process in wind-to-fuel initiatives involves capturing wind energy through wind turbines and using this energy to power electrolysis or other chemical processes. The electricity generated by the wind turbines is directed to electrolyzers, which split water into hydrogen and oxygen. This hydrogen can be used directly as a fuel or further processed into synthetic hydrocarbons through reactions with captured carbon dioxide.

Electrolysis: Proton Exchange Membrane (PEM) electrolyzers and Solid Oxide Electrolyzers (SOE) are commonly used in these projects. PEM electrolyzers are advantageous due to their high efficiency and fast response times, making them ideal for the variable nature of wind energy. Operating at high temperatures, SOE electrolyzers offer even higher efficiency and are suitable for large-scale industrial applications. Advanced power electronics systems manage the operation of these electrolyzers, ensuring optimal performance and energy efficiency.

Power Electronics: Power electronics are critical in wind-to-fuel systems, controlling the conversion and distribution of electricity generated by wind turbines. They optimize energy flow to the electrolyzers, maintaining stability and efficiency. Power electronics also enable the integration of energy storage solutions, ensuring a steady supply of electricity to the electrolyzers even when wind conditions fluctuate.

Table 8.1 comprehensively compares Proton Exchange Membrane (PEM) and Solid Oxide Electrolyzers (SOE) in the context of renewable-powered fuel production. It highlights key parameters such as operating temperature, efficiency, response time, energy input, and ideal applications. The table illustrates how PEM electrolyzers are best suited for dynamic renewable energy sources like solar and wind due to their rapid response time and low operating temperature, while SOE excels in industrial-scale applications, leveraging high-temperature operation and waste heat utilization to achieve superior efficiency. This side-by-side comparison aids in understanding the unique strengths and appropriate use cases for each technology.

8.6.4 Offshore Wind Hydrogen Production Plant

An exemplary project is an offshore wind hydrogen production plant. Wind turbines located offshore capture strong and consistent wind currents, generating electricity that powers PEM electrolyzers. The produced hydrogen is then transported to the shore for various applications, from fueling vehicles to generating electricity. This initiative demonstrates the viability of large-scale wind-to-fuel production, achieving high energy conversion efficiency and proving the concept's scalability. Figure 8.7 illustrates the setup of an offshore wind-to-fuel facility, showing wind turbines, electrolysis units, and synthetic fuel production processes. The figure illustrates an integrated offshore power generation system combining wind energy and gas turbine technology for efficient energy production. The wind farm supplies renewable energy to an offshore platform, an energy conversion and integration hub. Simultaneously, a gas turbine generates power, and its exhaust gases are utilized in a Waste Heat Recovery Unit (WHRU) to enhance system efficiency. The waste heat is directed to a Once-Through Steam Generator (OTSG), producing steam that drives a steam turbine for additional power generation. After passing through the turbine, the steam is condensed in a condenser to complete the cycle. This hybrid system combines

Table 8.1 Comparative analysis of PEM and solid oxide electrolyzers for renewable-powered fuel production

Parameter	PEM electrolyzers	Solid Oxide Electrolyzers (SOE)
Operating temperature	Low (20–80 °C)	High (700–1,000 °C)
Efficiency	Moderate (55–70%)	High (80–90%)
Response time	Rapid, suitable for dynamic renewable sources like solar/wind	Slower, optimized for steady industrial operations
Energy input	Electricity only	Electricity and thermal energy (can use waste heat)
Output	Hydrogen	Hydrogen or syngas (hydrogen and carbon monoxide)
Durability	Sensitive to impurities, requires high-purity water	More robust with respect to impurities
Cost	Higher due to expensive materials (e.g., platinum catalyst)	Lower material cost but higher thermal management expenses
Ideal applications	Integration with variable renewable energy sources	Industrial-scale applications with access to waste heat
Scalability	Well-suited for small to medium-scale systems	Better for large-scale and co-electrolysis processes

renewable wind energy with gas-based thermal power generation, maximizing energy output and improving efficiency through heat recovery.

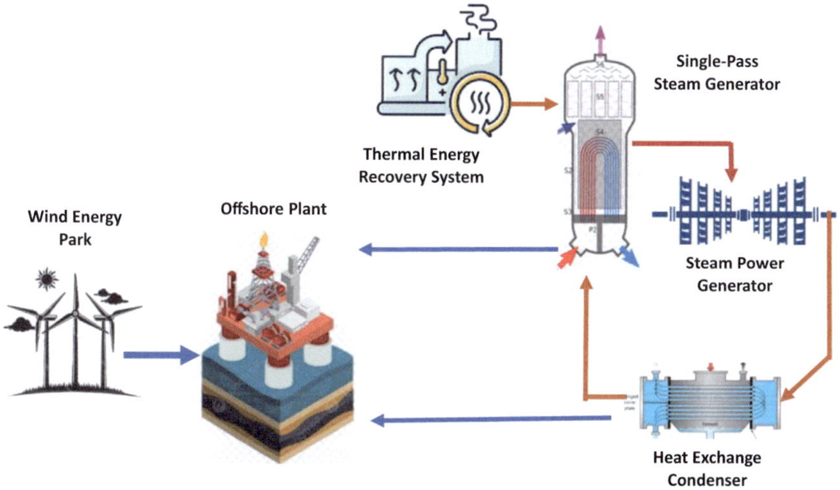

Fig. 8.7 Offshore hybrid power generation system: wind, gas turbine, and waste eat recovery integration

Wind-to-fuel initiatives offer numerous benefits, including reducing greenhouse gas emissions, enhancing energy security, and providing a sustainable fuel source. By utilizing wind energy, which is abundant and renewable, these projects contribute significantly to decarbonizing the energy sector. The wind-to-fuel initiatives represent a transformative approach to sustainable energy production. These projects convert wind energy into clean fuels by leveraging advanced power electronics and efficient electrolyzers, supporting the global shift toward a more sustainable and resilient energy future.

8.7 Future Directions

Future advancements in power electronics technologies hold the potential to further enhance the sustainability, scalability, and efficiency of e-fuel production. Emerging trends include the development of wide bandgap semiconductors, increasing integration of AI and machine learning for adaptive control, and exploring novel energy storage solutions. This chapter provides a visual overview of emerging trends in power electronics, such as wide bandgap semiconductors and AI-based control, and their potential impact on e-fuel production.

8.7.1 AI-Based Control Systems for Optimizing E-Fuel Production

AI-based control systems are pivotal in optimizing operational efficiency and scalability in e-fuel production facilities. These systems leverage machine learning algorithms to enable real-time adaptive management, dynamically predicting energy supply and demand based on historical data and real-time inputs. Such predictive capabilities allow for precise control of production processes, ensuring the optimal allocation of resources even under fluctuating renewable energy inputs.

AI intelligently coordinating plug-and-play modules enhances reliability and efficiency in modular power electronics systems. For instance, AI algorithms can detect and address inefficiencies or failures in individual modules, ensuring uninterrupted operations and seamless scaling of production capacity. This dynamic adaptability makes modular systems highly suitable for flexible and scalable e-fuel production facilities, capable of growing incrementally as demand increases.

Concrete examples of AI applications include its integration into Energy Management Systems (EMS), which optimizes the operation of renewable energy inputs, energy storage, and production processes in real-time. Similarly, AI-driven control in modular systems allows for automated scaling, minimizing downtime and reducing energy losses. These applications underscore the transformative potential of AI-based control in making e-fuel production systems more efficient, adaptive, and sustainable.

8.8 Conclusion

This chapter has highlighted the indispensable role of advanced power electronics in achieving a sustainable transport sector through efficient e-fuel production. Power electronics address critical challenges such as scalability, energy variability, and storage inefficiencies by integrating renewable energy sources, optimizing electrolysis, and employing modular and AI-enhanced systems. Case studies underscore the practical applications and potential of these technologies to enhance the resilience and sustainability of e-fuel production systems. As the global push toward decarbonization intensifies, the innovative use of power electronics will remain central to developing scalable and sustainable energy solutions. Future advancements in AI-based controls and wide bandgap semiconductors will further revolutionize e-fuel production, reinforcing its role in transitioning to a low-carbon transport sector.

References

Agarwal AK, Valera H (2022) Greener and scalable E-fuels for decarbonization of transport. Springer

Barbhuiya S, Kanavaris F, Das BB, Idrees M (2024) Decarbonising cement and concrete production: strategies, challenges and pathways for sustainable development. J Build Eng 86

Blaabjerg F, Yang Y, Ma K, Wang X (2015) Power electronics—the key technology for renewable energy system integration 1618–1626

Blaabjerg F, Yang Y, Kim K, Rodriguez J (2023) Power electronics technology for large-scale renewable energy generation. Proc IEEE

Chan KMA, Satterfield T (2020) The maturation of ecosystem services: social and policy research expands, but whither biophysically informed valuation. People Nat 2:1021–1060

Chen M, Chou S-F, Blaabjerg F, Davari P (1906) Overview of power electronic converter topologies enabling large-scale hydrogen production via water electrolysis. Appl Sci 2022:12

Dawood F (2020) Hydrogen production for energy: an overview. Int J Hydrogen Energy 45:3847–3869

Dell'Aversano S, Villante C, Gallucci K, Vanga G, Di Giuliano A (2024) E-fuels: a comprehensive review of the most promising technological alternatives towards an energy transition. Energies 17:3995

El-Shafie M, Kambara S, Hayakawa Y (2019) Hydrogen production technologies overview. J Power Energy Eng 7:107–154

Hafez FS, Sa'di B, Safa-Gamal M, Taufiq-Yap YH, Alrifaey M, Seyedmahmoudian M, Stojcevski A, Horan B, Mekhilef S (2023) Energy efficiency in sustainable buildings: a systematic review with taxonomy, challenges, motivations, methodological aspects, recommendations, and pathways for future research. Energy Strategy Rev 45

Kumar V, Mohanty SR (2023) Resilient optimal gain control and continuous twisting observer for enhanced power system performance under uncertainties. IEEE Syst J 17(2):2733–2744

Kumar V, Mohanty SR, Kumar S (2020) Event trigger super twisting sliding mode control for DC micro grid with matched/unmatched disturbance observer. IEEE Trans Smart Grid 11(5):3837–3849

Kumar V, Prasad U, Mohanty SR (2024) Entirely coupled recurrent neural network-based backstepping control for global stability of power system networks. IEEE Trans Autom Sci Eng 21(2):1647–1660

Kumar JCR, Majid MA (2020) Renewable energy for sustainable development in India: current status, future prospects, challenges, employment, and investment opportunities. Energ Sustain Soc 10(2)

Li H, Guo J, Li Z, Wang J (2023) Research progress of hydrogen production technology and related catalysts by electrolysis of water. Molecules 28(13):5010

Luo X, Wang J, Dooner M, Clarke J (2015) Overview of current development in electrical energy storage technologies and the application potential in power system operation. Appl Energy 137:511–536

Mahidin E, Husin H (2021) A critical review of the integration of renewable energy sources with various technologies. Prot Control Mod Power Syst 6(3)

Pessarrodona A, Franco-Santos RM, Wright LS, Vanderklift MA, Howard J, Pidgeon E, Wernberg T, Filbee-Dexter K (2023) Carbon sequestration and climate change mitigation using macroalgae: a state of knowledge review. Biol Rev 98:1945–1971

Popovic J, Ferreira JA (2010) Power electronics for sustainable energy future. Quant Value Power Electron 112–119

Prussi M, Lee U, Wang M, Malina R, Valin H, Taheripour F, Velarde C, Staples MD, Lonza L, Hileman JI (2021) CORSIA: the first internationally adopted approach to calculate life-cycle GHG emissions for aviation fuels. Renew Sustain Energy Rev 150

Reddy VJ, Hariram NP, Maity R, Ghazali MF, Kumarasamy S (2023) Sustainable E-fuels: green hydrogen, methanol and ammonia for carbon-neutral transportation. World Electr Veh J 14–349

Saeedmanesh A, Mac Kinnon MA, Brouwer J (2018) Hydrogen is essential for sustainability. Curr Opin Electrochem 12:166–181

Shahbaz M, Rashid N, Saleem J, Mackey H, McKay G, Al-Ansari T (2023) A review of waste management approaches to maximise sustainable value of waste from the oil and gas industry and potential for the State of Qatar. Fuel 332

Ueckerdt F, Bauer C, Dirnaichner A, Everall J, Sacchi R, Luderer G (2021) Potential and risks of hydrogen-based e-fuels in climate change mitigation. Nat Clim Chang 11:1–10

Wang H, Blaabjerg F (2020) Power electronics reliability: state of the art and outlook. IEEE J Emerg Select Top Power Electron 1–1

Chapter 9
Comprehensive Analysis of Electrofuels (E-Fuels) as Sustainable Fuels from Renewable Electricity: A Review

Suyash Wankhede, Jitendra Dixit, Dilip Sharma, and Nikhil Sharma

Abstract E-fuels are made from renewable power and collected carbon dioxide, have the potential to cut greenhouse gas emissions by up to 90% compared to traditional fuels. This review article investigates alternative e-fuel production pathways, including Power-to-Fuel (PtF) technologies, and evaluates their chemical processes, carbon capture methods, and nitrogen extraction strategies. Key e-fuels such as e-methane, e-methanol, and e-ammonia are assessed for their manufacturing processes and possible applications. The review also looks at the life cycle impacts, production costs, and combustion properties of e-fuels. Scalability challenges, infrastructure development, and cost competitiveness have been cited as significant factors impacting mainstream adoption. The article discusses about successful integration of e-fuels in the global energy landscape is hinged on technological advancements, supportive regulatory frameworks and economic incentives. This review adds to an ongoing debate about sustainable energy options by offering some insights into the potential role that e-fuels could play in addressing climate change and reducing dependence on fossil fuels.

Keywords E-fuels · Electrofuels · E-Methane · E-Methanol · E-Ammonia · Combustion · Chemistry of e-fuels

9.1 Introduction

Climate change poses a serious threat to our world, and the transportation industry is a major source of greenhouse gas emissions. Gasoline and diesel are still prevalent in this sector thus emitting a lot of carbon dioxide (CO_2) into the atmosphere. To attain the ambitious targets set out in global pacts like Paris Agreement, a move from non-sustainable transport fuels to sustainable fuel sources is necessary. Modern society is

S. Wankhede · J. Dixit · D. Sharma · N. Sharma (✉)
Department of Mechanical Engineering, Malaviya National Institute of Technology Jaipur, Jaipur, Rajasthan, India
e-mail: nikhil.mech@mnit.ac.in

N. Sharma et al. (eds.), *Sustainable, Scalable and Storable E-Fuels for Decarbonising Transport Sector*, Energy, Environment, and Sustainability,
https://doi.org/10.1007/978-981-96-7388-9_9

in continuous search of solutions for reducing Greenhouse Gases (GHG) emissions, especially in the industrial and transport sectors, in a sustainable way. These two sectors contributed in about 75% of CO_2 emissions in the last decade as appraised in a recent works (Change 2014). The overall global energy usage is about 20 TW with the approximate CO_2 concentrations of about 400 ppm. Since, more than half of the world's population is below the poverty line and access to power has a significant impact on poverty reduction. Above the poverty line creates an extra 10 TW energy while emitting no CO_2.

In this context electrification has the ability to reduce greenhouse gas emissions from the transport sectors, industrial sectors and construction sectors, also depending on the resources utilized for power generation. For instance, the US generates almost 70% of its power from almost zero or low-carbon emitting fuels, resulting in little CO_2 emissions compared to fossil fuels (Cleary 2019). The usage of electricity as a source of energy for powering the vehicles causes less carbon emissions as compared to the traditional fossil fuels. Electrification is predicted to continue to grow in transportation, buildings, and industry. However, electrification is not feasible for certain applications including aircraft, large load vehicles and maritime transport due the requirement of high energy and power and also due to low volumetric density of the batteries. To overcome these limits for e-fuels a new term was introduced Power-to-Fuel (PtF). The goal of PtF is to transform power into carbon–neutral synthetic fuels that can be used to replace fossil fuel products. The term power refers to the energy which gets generated while utilizing the renewable energy sources. The term fuel refers to a variety of synthetic hydrocarbon (HC) fuels, i.e. Power-to-Gas, Power-to-Liquid and Power-to-Fuel. Such type of synthetic HC fuels is known as e-fuels or electrofuels.

The most important gas for the Power-to-Fuel is hydrogen gas, i.e. power-to-hydrogen, i.e. converting the hydrogen gas into a synthetic fuel. The PtF can produce various synthetic fuels derived from the hydrogen gas, such as methane, methanol, ammonia, with a relatively low capital intensive decarbonization approach to reduce CO_2 emissions from energy sector (Daiyan et al. 2020). The synthetic fuel or e-fuel is a type of energy carrier in which the energy is stored between the chemical bonds of the fuels molecules (Tarascon and Armand 2001) the stored energy for petrol fuel is 46 MJ/kg (Levine et al. 2010).

The terms e-fuel and synthetic fuel were used interchangeably many times in the literature, The synthetic fuels are those which are produced from renewable sources via thermochemical conversion process to a gas or liquid fuel (Ridjan et al. 2016). There are two primary ways to manufacture electrofuel: the first is conversion of CO to fuels which has high energy density (Hawkins et al. 2013) and the second is combining CO_2 gas emitted from sources and the hydrogen gas obtained via electrolysis of water (Pearson and Turner 2012). Figure 9.1 depicts energy conversion pathways from solar energy to biodiesel (Mehrpooya et al. 2020).

Metal fuels are created by electrochemical reduction processes of their oxides; if electricity was used as the reaction's energy source, these fuels can be referred to as electrofuels. This terminology is sometimes used in literature. The obtained metal fuel can then be used in according to the user's needs in the metal fuel cycle

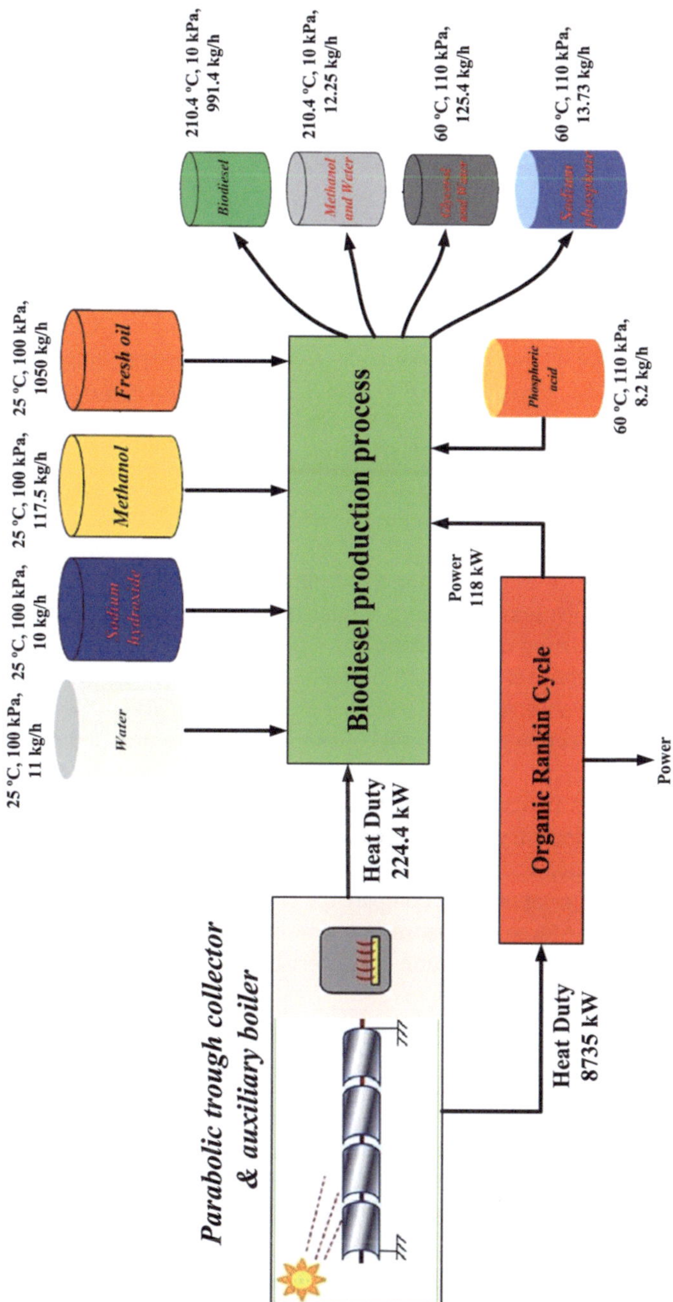

Fig. 9.1 Block flow diagram of the integrated system of biodiesel production process (Mehrpooya et al. 2020)

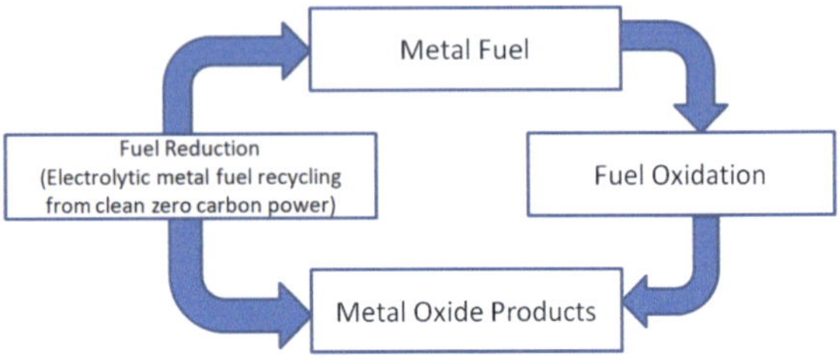

Fig. 9.2 Typical metal fuel redox cycle (Julien and Bergthorson 2017)

(Fig. 9.2), there the metal fuel gets burned (oxidized) in an atmospheric air combustor for generating power.

The word "electrofuel" in this review refers to gas or liquid fuels that primarily employ carbon dioxide as a carbon source, as well as hydrogen produced by the electrolysis of water. The energy needed to manufacture these fuels is derived from zero-carbon electric energy. Ammonia made from clean energy sources and renewable carbon sources is one of the liquid fuels. Similarly, solid metal fuel that has been reduced with clean, zero-carbon electricity will also be included in the phrase. The e-fuels play a major role in decarbonizing the energy sectors as these are carbon–neutral synthetic fuels which also helps in reducing the greenhouse gas emissions, e-fuels enables the storage and utilization of surplus renewable energy and they can be used in various applications, including combustion engines and fuel cells with a little modification in existing infrastructure.

In an article by Brynolf et al. (2018) a number of methods were discussed to evaluate the production costs for e-fuels. One of the methods analyses the costs and efficiencies of the production cost of e-fuels from various literatures and compares them.

The objective of this review is to understand the methodology required for the e-fuels including the chemistry of the e-fuels, provided with the chemical reactions to prove it. We also discuss about the carbon capture and storage (CSS) technologies and the source for carbon capture and compare the electrofuels life cycle assessment from different studies. Moreover, this review also covers the factors affecting the cost of e-fuels and also their combustion characteristics and future prospects.

9.2 Methodology

The methodology required for the production of synthetic fuels or e-fuels from renewable power sources such as electricity is discussed in this section. Figure 9.3 presents a schematic of a general overview of renewable Power-to-Fuel technology. Selecting an optimum place for an e-fuel production plant involves a complex interplay of various factors such as its proximity and transportation, the plant should be as close as possible to its target markets which minimizes the cost for transportation. The availability of energy sources is also an important factor for finalizing the location of plant.

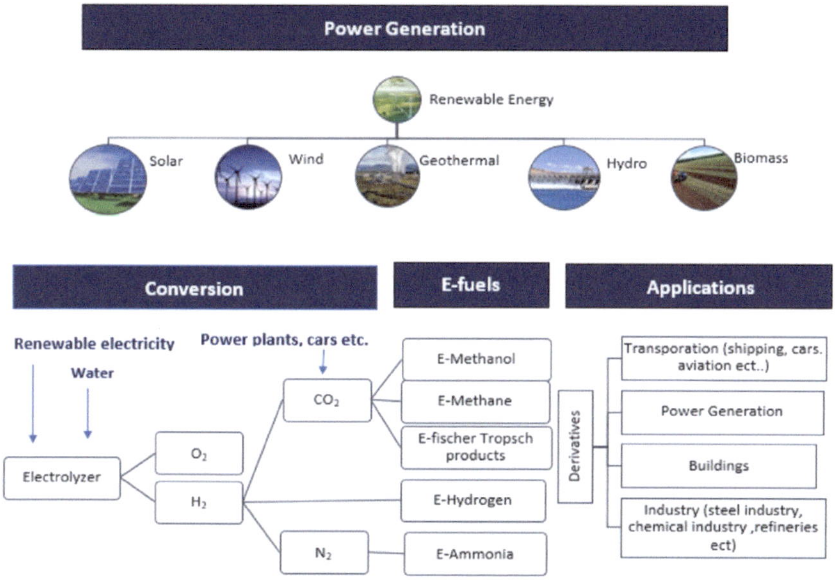

Fig. 9.3 A schematic of power-to-fuel process (Nemmour et al. 2023)

Therefore, in order to take the appropriate safety precautions that have an impact on the plant's performance, it is important to investigate the location's physical attributes before the designing and operations phase. In addition, the economics of renewable energy sources must be taken into account together with their infrastructure.

The power-to-hydrogen is a process in which the electricity obtained from the renewable source is utilized for generating hydrogen via water electrolysis. When power is produced using renewable resources, electrolysis is crucial to the production of renewable hydrogen. Although this method is well-known and devoid of greenhouse gases, it accounts for around 4% of the total hydrogen production (Nicita et al. 2020). This is due to the lack of infrastructure such as transport and also problems related to the electrolysis production stage that includes high production cost, energy losses during process and sustainability concerns.

9.2.1 *Electrolysis*

Water electrolysis is the process of applying electrical energy to split water into hydrogen and oxygen (reaction 9.1). The following lists the electrochemical reactions that take place at the electrodes of an alkaline electrolyzer (Rodríguez and Amores 2020). The reactions 9.2 and 9.3 depict the reactions taking place at the cathode and anode, respectively.

$$\text{Overall:}H_2O(l) \rightarrow H_2(g) + \frac{1}{2}O_2(g) \tag{9.1}$$

$$\text{Cathode:}2H_2O + 2e^- \rightarrow 2OH^- + H_2 \tag{9.2}$$

$$\text{Anode:}2OH^- \rightarrow \frac{1}{2}O_2 + H_2O + 2e^- \tag{9.3}$$

The water electrolyzers are of three types mainly which are alkaline electrolyzers (AEC), polymer electrolyte membrane (PEM), and solid oxide electrolyzers (SOE).

9.2.1.1 Alkaline Electrolyzers (AEC)

Alkaline electrolyzer cell is referred to as AEC. This type of electrolyzer is used in the process of water electrolysis, which uses electricity to separate water molecules (H_2O) into hydrogen (H_2) and oxygen. It is the most advanced commercially available technology and has been widely used in the chemical industry for large-scale hydrogen synthesis. AECs, like other electrolyzers, separate water molecules by sending an electric current through an electrolyte solution. Strong alkaline solutions, such as potassium hydroxide (KOH) or sodium hydroxide (NaOH), are frequently employed as electrolytes in AECs. The electrolyte solution is immersed between two electrodes, a cathode and an anode. The cathode generates hydrogen gas, whereas the anode changes oxygen gas as energy travels through the system. The operating temperature range is at 60–80 °C. The AEC electrolyzers operates at either atmospheric pressure or at a higher-pressure range. Pressurized alkaline electrolyzers has the advantage of producing compressed hydrogen, which can be utilized in other applications or immediately pumped into grids without needing to be compressed, i.e. less energy is consumed and no additional energy input is needed. The disadvantages are that the pressurized AEC has lower efficiency as compared to the atmospheric AEC with low purity of produced hydrogen and AEC also has high maintenance cost as it is associated with highly corrosive electrolytes (Götz et al. 2016).

When large-scale hydrogen production is needed for commercial processes like chemical manufacture or ammonia synthesis, AECs are frequently employed. Despite their reduced efficiency, they may be appropriate for some applications due to their simpler design and cheaper initial cost. Future developments in the clean energy sector could see the use of AEC technology as investigations into enhancing efficiency and resolving constraints continue. The Fig. 9.4 describes the operating principle of the alkaline water electrolysis.

9.2.1.2 Polymer Electrolyte Membrane (PEM)

Proton exchange membrane fuel cells (PEMFCs) and proton exchange membrane electrolyzers (PEM electrolyzers) are two examples of electrochemical devices

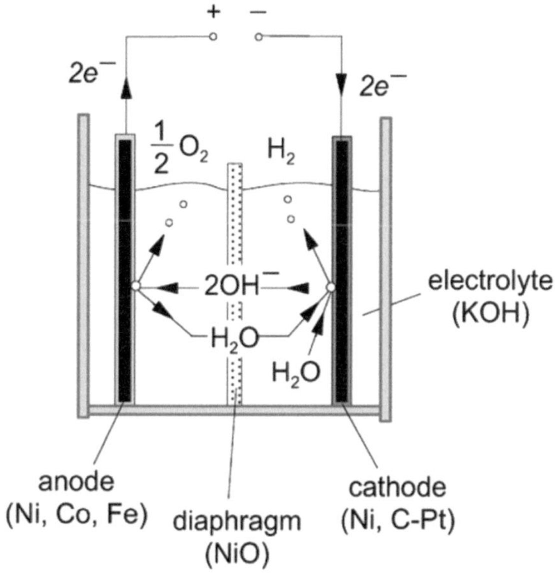

Fig. 9.4 Operation principle of alkaline water electrolysis (Ulleberg 2003)

that depend on polymer electrolyte membranes (PEMs), also referred to as proton exchange membranes. Polymer electrolyte membranes (PEMs) are essentially thin sheets of plastic that have been modified to allow them to conduct protons (H+) while blocking electrons and reactant gases. This unique combination of properties makes them ideal for separating the anode and cathode reactions in electrochemical devices while still allowing the efficient transfer of protons between them. Solid polymer electrolyte is used in PEM electrolyzers with an operating temperature not more than 80 °C.

The operating principle of PEM water electrolyzer is depicted in Fig. 9.5. In the PEM water electrolyzer, the electrochemical cell reactions occur at both electrodes when a voltage larger than the thermos-neutral potential for water splitting is applied on it. The reaction rate for water splitting is determined by the applied potential, temperature, anode and cathode electrocatalyst characteristics and membrane resistance (Aricò et al. 2013).

The PEM electrolyzers are distinguished by high purity and low-density electrolysis (Götz et al. 2016). Since PEM uses noble meal as catalysts its capital cost is more as compared to AEC. AEC electrolyzers are known for their enormous production capacity and great efficiency when it comes to producing hydrogen, while PEM electrolyzers are known for their lower production capacity and average efficiency.

9.2.1.3 Solid Oxide Electrolyzers (SOE)

The SOE electrolyzer is one of the most recent technologies which is still in under-development stage. Two porous electrodes and a dense ionic conducting electrolyte

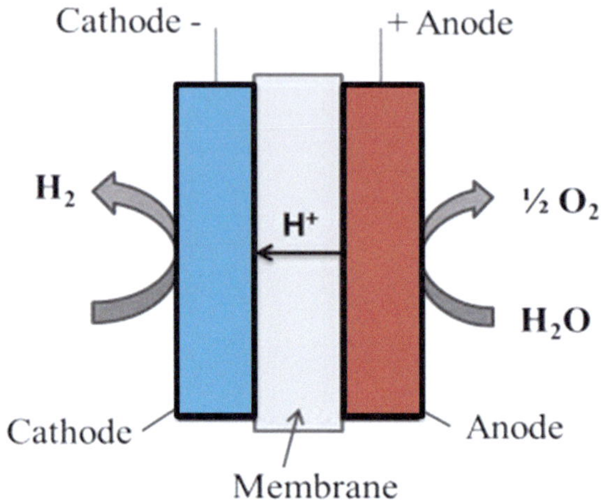

Fig. 9.5 Operating principle of PEM water electrolyzer (Aricò et al. 2013)

are the essential parts of an SOEC. Figure 9.6 illustrates the basic mechanics of SOEC operation. Steam is delivered to the porous cathode. Water molecules diffuse to the reaction sites and dissociate to form hydrogen gas and oxygen ions at the cathode–electrolyte interface when the necessary electrical voltage is applied to the SOEC. The generated hydrogen gas diffuses and gathers at the cathode surface. The thick electrolyte transports the oxygen ions to the anode. The oxygen ions on the anode side undergo oxidation to form oxygen gas, which is then carried to the anode surface via the anode's pores (Ni et al. 2008). The net reaction of SOEC is:

$$H_2O \rightarrow H_2 + 1/2\,O_2$$

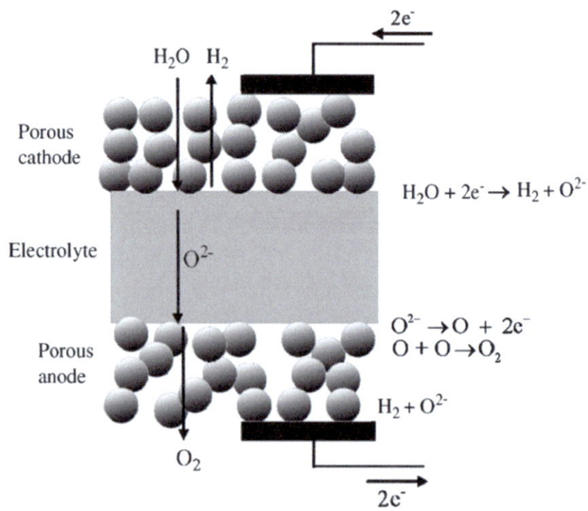

Fig. 9.6 Schematics of SOEC hydrogen production (Ni et al. 2008)

The SOE electrolyzers has high-temperature ranges as it uses oxygen ions conducting cells. This electrolyzer has excellent thermal and chemical stability as well as increased energy efficiency. Additionally, because of its excellent energy conversion efficiency, it uses less electricity. Nevertheless, the cells' limited long-term stability and rapid material deterioration are the outcomes of the high-temperature operating range (Ziabari and Ghandehariun 2020). The challenges faced during SOE includes maintaining high-temperatures requires extra energy and specialized materials.

9.3 Chemistry of E-Fuels

Hydrogen is regarded as the perfect electrofuel and is anticipated to serve as both an intermediary in the synthesis of fuels and value-added chemicals and the primary energy source of future energy systems. But hydrogen has two drawbacks: it's hard to store and has a poor volumetric energy density. One of the primary methods for producing hydrogen is water electrolysis. Its primary benefit is its excellent energy efficiency, up to 80%.

9.3.1 E-Methane

Methane is one of the fuels that can be readily found in nature in its original form and it is a major constituent of natural gas. Despite being one of the most significant greenhouse gases, it is used in many equipment's in the houses as a fuel, water heating, ovens, kilns, cars and other equipment's. Synthetic natural gas (SNG) can be injected directly into the natural gas grid, taking advantage of existing natural gas infrastructure (Götz et al 2016). Power-to-methane (PtM) technology has proven to be a potential method for sustainable methane production, utilizing CO_2 and renewable energy to generate SNG. Among the various extant PtM methods, catalytic CO_2 hydrogenation (methanation) has received substantial research, and demonstration units are already operational in several countries.

9.3.1.1 Production Process

Methanation and electrolysis can be coupled in two processes to yield e-methane. Initially, water electrolysis employing renewable energy sources such as solar, hydro and wind power generates H_2. Following the absorption of CO_2 by a CCS plant, e-methane is produced by hydrogenating the gas in a methanation reactor using the Sabatier process.

$$4H_2 + CO_2 \rightarrow CH_4 + 2H_2O \ \Delta H_r = -165\ \text{kJ/mol}$$

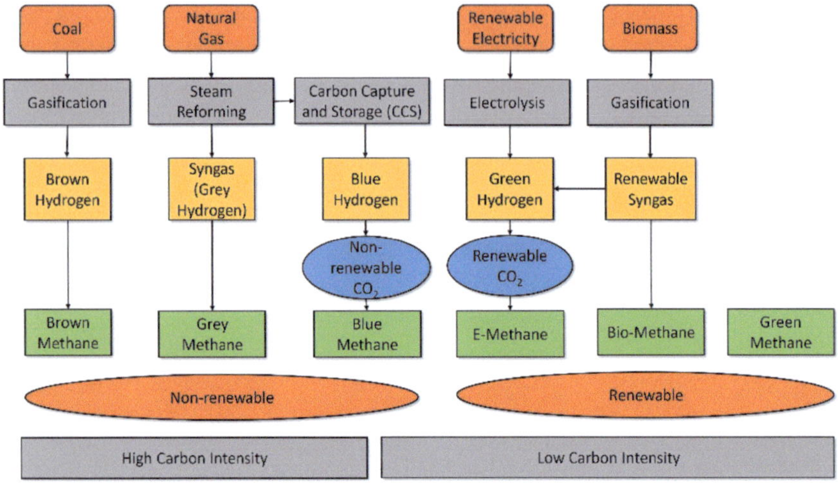

Fig. 9.7 Methane production pathways (Nemmour et al. 2023)

This chemical reaction is a highly exothermic reaction, i.e. it releases heat energy. Although this is a fast process it requires large amounts of CO_2 which can be difficult to obtain as the carbon capture and storage facilities are located at places far away from the plants thereby increasing the transportation expense for CO_2.

The Fig. 9.7 depicts the several paths of methane generation. The production cost for methane is lower as compared to the cost of production for fossil fuels (grey and brown methane). The methanation method is the most prominent method used for the power-to-methane projects in various countries (Rego de Vasconcelos and Lavoie 2019). From Fig. 9.7 we can observe that from all the methane production pathways the power-to-e-methane process, i.e. obtaining e-methane via water electrolysis and renewable power source (electricity) is renewable and has low-carbon intensity.

9.3.2 *E-Methanol*

E-methanol, a versatile, clean-burning fuel derived from renewable energy sources, is gaining popularity as a crucial player in the fight against climate change. It can store renewable energy in liquid form, making it portable and useful for a variety of applications such as transportation (Stančin et al. 2020), power generation. It also acts as a valuable feedstock for the chemical industry, creating critical items such as plastics, paints and cosmetics. Reducing emissions from the manufacture of methanol and developing alternative methods for producing renewable methanol are essential steps toward decarbonizing the chemical sector.

9.3.2.1 Production Process

Carbon dioxide is converted to methanol by a catalytic hydrogenative process, as follows:

$$CO_2 + 3H_2 \rightarrow CH_3OH + H_2O\ \Delta H_r = -49.2\ \text{kJ/mol}$$

The above chemical reaction is an exothermic reaction, i.e. it releases heat energy during the process with an operating temperature of 473–573 k and an CO_2:H_2 ratio of 1:3 in the presence of the copper/zinc oxide-based catalyst (Rivarolo et al. 2016). Methanol can be produced from a variety of carbonaceous sources, such as natural gas, coal, biomass, byproducts and CO_2 obtained from the industrial emissions via direct air capture. The Fig. 9.8 outlines a plan for producing methanol from different feedstocks.

It can be noticed from the Fig. 9.8 that green methanol, made from renewable electricity and collected CO_2, has the lowest carbon footprint. The carbon impact of e-methanol made from green hydrogen and renewable CO_2 through the use of renewable energy sources is minimal. Grey methanol is produced from natural gas without carbon capture, whereas blue methanol uses natural gas with carbon capture and storage, resulting in lower emissions. With regard to carbon intensity, brown methanol made from coal has the greatest. In addition to direct air capture (DAC), blue hydrogen can be injected into the methanol synthesis process to produce blue methanol, which is another method of lowering CO_2 emissions (Noussan et al. 2020).

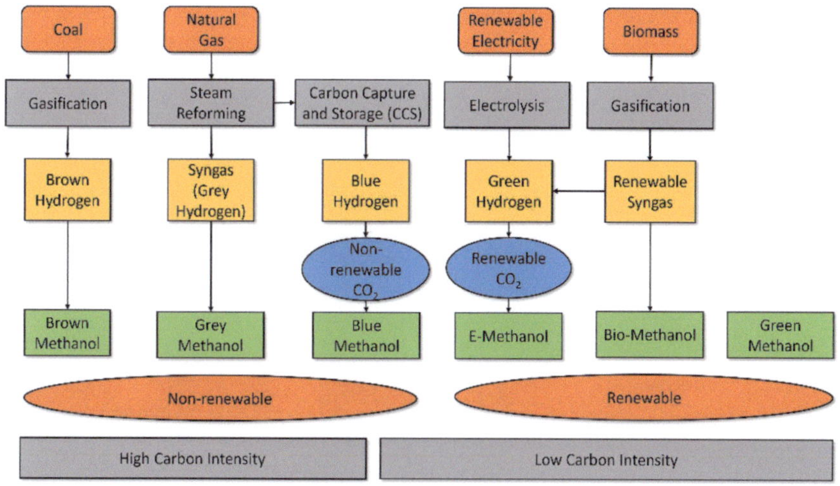

Fig. 9.8 Methanol production pathways (Kang 2021)

9.3.3 E-Ammonia

Ammonia (NH_3) is a chemical molecule which is trending in recent times for its potential to be used as an alternative fuel (Stančin et al. 2020). Ammonia is used for a variety of purposes, although agricultural fertilizers account for approximately 80% of annual production. It also plays an important role in the production of explosives, polymers, textiles, refrigerants and cleaning agents. Ammonia is also used to produce nitric acid and soda ash. Ammonia can also be used as a synthetic fuel in internal combustion engines for renewable energy.

9.3.3.1 Production Process

Ammonia is created in two ways: the Haber–Bosch process and solid-state ammonia synthesis (SSAS) (Fig. 9.9). The below equation depicts the Haber–Bosch process of generating ammonia. The green ammonia is produced in a similar manner as Haber–Bosch process, here the hydrogen used is obtained via water electrolysis using renewable energy source and the nitrogen is separated from air.

$$3H_2 + N_2 \rightarrow 2NH_3 \Delta H_r = -91.8\ \mathrm{kJ/mol}$$

Figure 9.10 shows the current production routes for ammonia. The primary sources of ammonia are fossil fuels like coal and natural gas. The ammonia that is obtained in this way is called brown ammonia, whereas the ammonia that is created via CCS from fossil fuels is known as blue ammonia.

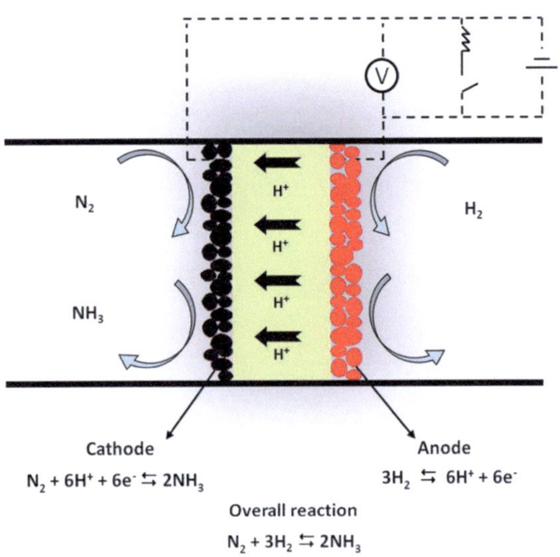

Fig. 9.9 Principle of operation of SSAS

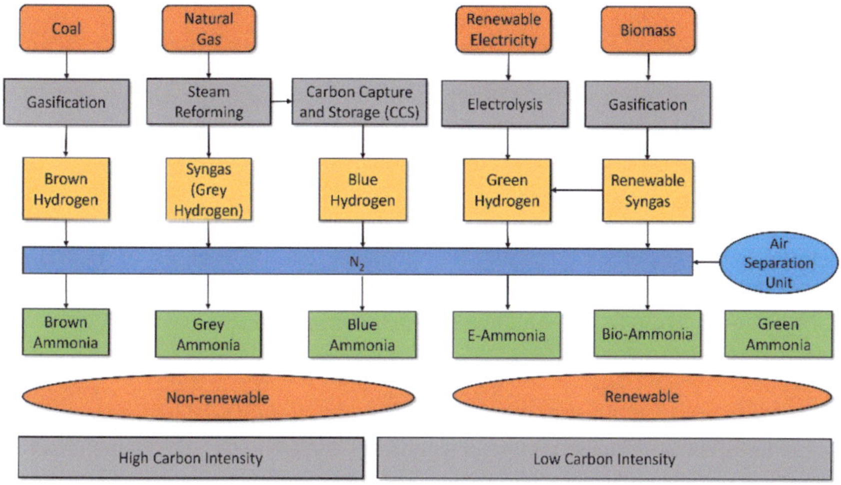

Fig. 9.10 Ammonia production pathways (D'Alessandro et al. 2010)

9.4 Carbon Capture and Storage (CCS)

The technique of carbon capture and storage (CCS), involves capturing CO_2 that has been generated and storing it in large-scale facilities for various further applications. The CCS is an important system which is closely connected to the e-fuels production method as the captured CO_2 serves as the primary feedstock for the e-fuels production, the CCS technologies ensure a steady and purified supply of CO_2 which is critical for the efficiency of e-fuels synthesis. CCS is a crucial method for lowering climate change because it primarily tackles atmospheric carbon dioxide, which is the greenhouse gas. The method of capturing CO_2 emissions comes from industrial sources including power plants by CCS in minimizing global warming. In addition, the captured CO_2 is safely stored within geological formations, barring it from recirculation into the atmosphere. This technique provides significant help in moderating greenhouse gas levels, especially in sectors that struggle with decarbonisation problems. Apart from this, when used along with bioenergy with carbon capture and storage (BECCS), it can potentially lead to negative emissions due to active removal of CO_2 from the environment. Even though certain technical and financial challenges persist, CCS holds value as an effective climate change mitigation strategy warranting further investment in research and development. The process of CCS encompasses three steps, capturing CO_2, its compression and transportation for permanent storage. Both natural processes and human activities are predominantly responsible for CO_2 production.

9.4.1 Sources for CO_2

There are various sources of CO_2, which are critical for understanding the CCS technologies. Given below are some of the primary CO_2 sources in detail.

1. Fossil Fuel Combustion

The majority of anthropogenic CO_2 emissions come from power plants, industrial facilities and transportation sectors. Power plants that use coal, oil or natural gas emit high levels of CO_2 due to their carbon content. Burning fossil fuels for energy and heat contributes significantly to global CO_2 emissions.

2. Industrial Processes

Several industrial activities emit CO_2 as a byproduct. Cement production, steel manufacture and chemical industries are all significant contributors. Cement manufacture releases it during limestone calcination. The steel industry produces CO_2 during the reduction of iron ore in blast furnaces. Chemical activities, such as producing ammonia and hydrogen, contribute to industrial CO_2 emissions.

3. Natural Gas Processing

Natural gas extraction and processing need eliminating CO_2 and other contaminants to fulfil quality standards. The method produces CO_2 as a waste product. Furthermore, methane (CH_4) leaks during extraction and processing can oxidize to CO_2 in the atmosphere, increasing the total emissions from this industry.

4. Biomass Decomposition and Combustion

Decomposing organic stuff, including plant material (Trost et al. 2016), naturally releases CO_2. However, using biomass as a fuel source, whether through direct combustion or bioenergy production, also emits CO_2. Despite being carbon–neutral under specific conditions (when CO_2 released is about equivalent to CO_2 absorbed during the plant's lifecycle), this source nevertheless contributes to the carbon cycle (Arasto et al. 2014).

5. Waste Management

Landfills and trash incineration generate CO_2 emissions. Organic waste in landfills decomposes anaerobically, creating methane that eventually oxidizes into CO_2. When organic materials are burned during waste incineration, CO_2 is directly released.

6. Agriculture

Agricultural methods include soil management, rice farming and livestock raising contribute to CO_2 emissions. Soil respiration and organic matter breakdown contribute to CO_2 emission. Furthermore, agricultural machinery and fertilizer use provide an indirect contribution through fossil fuel combustion and industrial processes.

9.4.2 Technologies

Direct CO_2 capture is possible with both liquid and solid direct air capture (DAC) technologies. Adsorbents used in Solid DAC (S-DAC) function at medium temperature (80–120 °C) and low pressure. An aqueous basic solution, like potassium hydroxide, is used by Liquid DAC (L-DAC) to release collected CO_2 through high-temperature units (300–900 °C). Existing CCS technology can handle 85–95% of a power plant's produced CO_2, reducing current emissions by around 80–90% (Mathieu 2006; Wilberforce et al. 2021). The process of integrating these systems in power plants requires a lot of energy, with an extra 10–40% needed for CO_2 capture (Mathieu 2006). The amine-scrubbing method is the most widely used approach for CO_2 extraction. Still, it's in operation, and the maintenance costs are high. As a result, new technologies are developing and becoming more and more popular because of their many advantages. Examples of these technologies are membranes and low-temperature technologies. However, due to constraints, these approaches cannot capture all (50–94%) emitted CO_2.

The carbon capture time of the burning of fossil fuels is the basis for classifying the present carbon capture and storage methods (Wilberforce et al. 2021; D'Alessandro et al. 2010), they are post-combustion, pre-combustion and oxyfuel combustion. Post-combustion capture (Fig. 9.11) removes carbon dioxide (CO_2) from exhaust gases after fossil fuel combustion, making it ideal for retrofitting existing power plants and industrial sites. The most commercially advanced technique for this technology is wet scrubbing with aqueous amine solutions. This procedure involves passing CO_2-containing flue gas through a scrubber and interacting with an amine solution at roughly 50 °C to generate a chemical. The CO_2-laden solution is heated to 120 °C in a regenerator, releasing the CO_2. It is then captured, dried, compressed and sent to geological storage. Post-combustion capture is a rapid and versatile option for reducing CO_2 emissions. It can work with many fuel types and integrate with existing power generation systems without disturbing main energy output.

However, the system has several challenges, including high energy requirements for solvent regeneration, which can reduce overall power plant efficiency, as well as running costs for solvent degradation and replenishment. Environmental concerns around the consequences of amine degradation products must also be addressed. Ongoing research aims to improve solvent efficiency and stability, optimize capture and regeneration processes, and better integrate capture systems with power plants in order to reduce energy losses. Advances in post-combustion capture make it a cost-effective solution for lowering CO_2 emissions on a large scale (Gibbins and Chalmers 2008).

In the pre-combustion technique (Fig. 9.12), the carbon dioxide is captured or absorbed before the combustion takes place, the fossil fuels are gasified at high pressures (30–70 atm) to produce syngas, i.e. a mixture of CO_2 and H_2. The syngas undergoes the water–gas shift process, which converts CO and water into CO_2 and H_2. CO_2 is separated from the hydrogen-rich gas using a physical solvent. The solvent

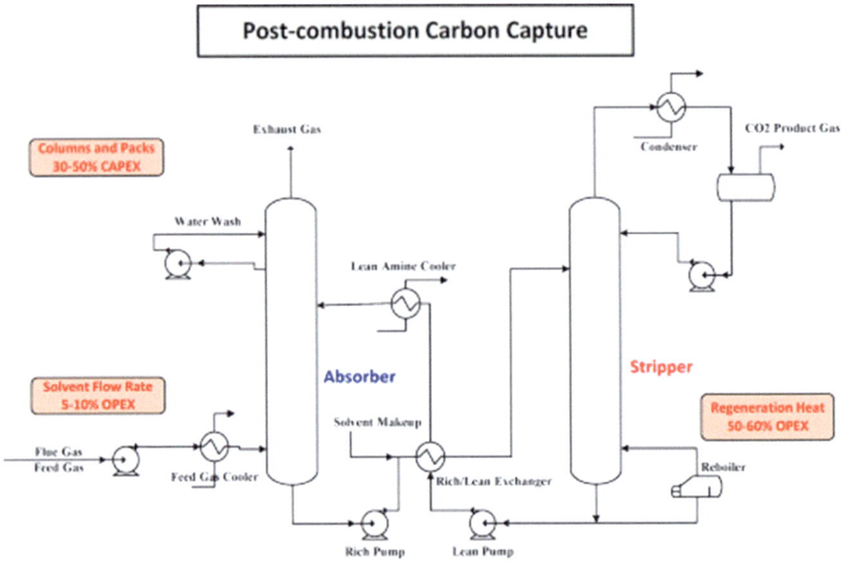

Fig. 9.11 Post-combustion carbon capture process (Liang et al. 2015)

absorbs CO_2 at high pressure and releases it when pressure is reduced, reducing energy needs compared to post-combustion procedures.

This technique efficiently captures and compresses CO_2, but has efficiency penalty. The shift reaction reduces the mass accessible for burning, lowering the power generation potential. Furthermore, turbines that use hydrogen-rich fuel work less effectively due to increased heat transfer coefficients, necessitating lower turbine inlet temperatures. Gasification efficiency losses occur for both solid and liquid fuels, particularly when heated products are cooled by water sprays, which add water to the shift process but diminish overall efficiency. Despite these challenges, pre-combustion capture is advantageous due to lower energy requirements for CO_2 capture and compression. However, the trade-offs in efficiency must be considered

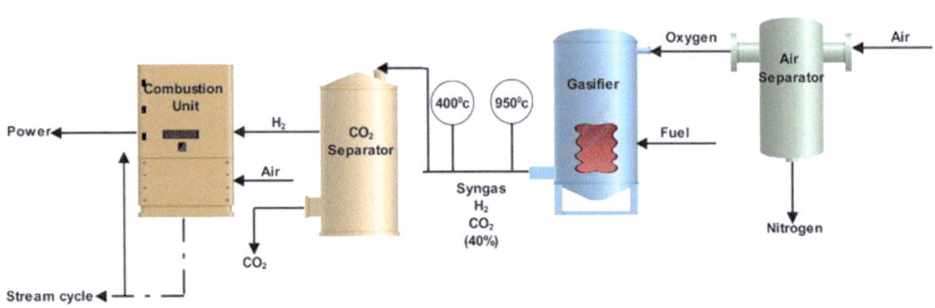

Fig. 9.12 Schematics of the pre-combustion carbon capture technique (Jimoh et al. 2023)

when evaluating its overall effectiveness and applicability (Gibbins and Chalmers 2008).

Oxyfuel combustion (Fig. 9.13) burns the fuel (coal or gas) in a combination of oxygen and recycled flue gases, rather than in air. The primary separation stage is to remove oxygen from nitrogen in the air before combustion. This is typically accomplished through cryogenic distillation. The fuel is then burned in an oxygen mixture with recovered flue gasses. The recycled flue gases replace the nitrogen that would otherwise be present in air combustion. This helps to limit peak flame temperatures in boiler designs, taking into account material and ash-slagging constraints.

The resultant flue gas is composed primarily of CO_2 and condensable water vapour. This makes it reasonably simple to separate and purify the CO_2 during the compression cycle. Pollutants such as Nitrogen oxide (NOx), Sulphur oxide (SOx) and other impurities must be removed from coal plants' product gas prior to or during CO_2 compression. Current oxyfuel combustion designs for natural gas plants that use cryogenic oxygen production appear uncompetitive, even before accounting for the expenses associated with developing new gas turbine designs. However, for coal plants, oxyfuel combustion has the potential to be more efficient and cost-effective than amine-based post-combustion capture.

One significant potential advancement for oxyfuel systems is the use of ion transfer membranes for oxygen synthesis rather than cryogenic distillation. Another option is to employ very small high-temperature "rocket burner" oxyfuel steam turbine systems that use water recycling rather than flue gas recycling to achieve moderate combustion temperatures. Overall, oxyfuel combustion is seen as a viable CO_2 capture method for coal plants, with various utilities and manufacturers attempting to develop and demonstrate the technology. However, when compared to typical air-fired facilities, new equipment designs are still required (Liang et al. 2015).

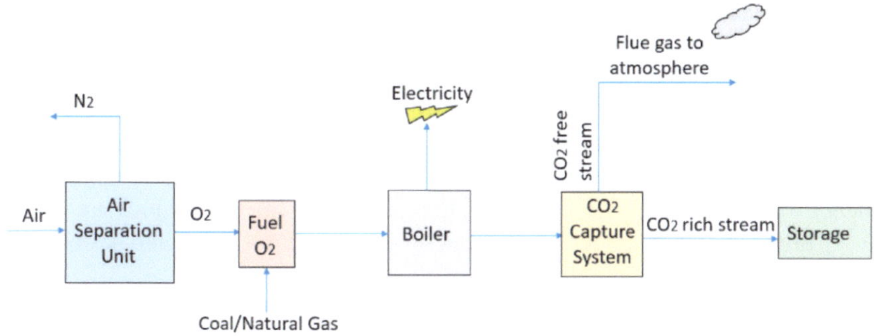

Fig. 9.13 Block flow diagram of an oxyfuel combustion carbon capture process (Elhenawy et al. 2020)

9.5 Nitrogen Capture

While carbon capture has a rather limited application, nitrogen capture offers a more comprehensive plan that serves a purpose in many environmental and industrial applications. Though it isn't focused on climate change, results could be positive in farming and industrial processes regarding sustainability. Extracted nitrogen gas offsets the application of the vigorous and power-loving Haber–Bosch process that is used to produce fertilizer with the strong mark of agriculture on environment. Captured nitrogen has the potential to be used in a variety of industrial applications, such as creating inert atmospheres or as a component in electronics manufacture. In addition to these practical uses, nitrogen capture serves to aid in the treatment of wastewater by lessening the ammonia levels in water discharge and minimizing the detrimental effects of nitrogen leaching in the environment. In the future using this technique it may be possible to harvest nitrogen on nitrogen bearing worlds which support life and enable transport vehicles or space ships. The continued development of large-scale nitrogen capture technologies holds great potential for a more sustainable future in many areas. The captured nitrogen plays a vital role in ammonia production, there are three methods that are used commercially for nitrogen generation from air which are cryogenic distillation, membrane separation and pressure swing adsorption (PSA).

To separate nitrogen from atmospheric air, cryogenic distillation is utilized at very low temperatures. Initially, the air is filtered and compressed to eliminate pollutants and increase its pressure. Then this compressed air is cooled utilizing a series of heat exchangers to cryogenic temperatures. The cooled air enters into a high-pressure distillation column for the first separation process. Nitrogen being more volatile ascends on the top of the column while oxygen and other less volatile gases settle at the bottom. The nitrogen-rich vapour then passes through another low-pressure distillation column for further purification. In this case, the column uses multiple VR/condensers that enhance splitting efficiency by condensing nitrogen at different levels. By doing so it reduces exergy losses as well as improving overall effectiveness hence addressing disadvantages of conventional single-column systems. They allow efficient heat transfer in addition to reducing extra boil-up necessary; thereby saving energy used in these facilities. Exergy analysis identifies and helps in minimizing energy wastage during process optimization. Efficiency increases which are particularly important when producing nitrogen on an industrial scale should be achieved by maximizing such processes (Guerra et al. 2020).

For nitrogen separation from air, membrane separation acts by separating gases through their permeation rates. This approach exploits the difference in the rate of diffusion of oxygen and nitrogen across a membrane due to an affinity and molecular size variations. Smaller sized oxygen molecules diffuse faster through the membrane, thus leaving behind a feed side that is rich in nitrogen. Simplicity and safety are attributes of this method because it has got a direct design with less risk involved during operation procedures. The drawback about this process is that it often fails to achieve sufficient efficiency in producing high purity nitrogen which is quite limiting. A liquefaction unit usually becomes mandatory for production of higher purity nitrogen. However, within these limits, membrane separation still remains as a viable alternative particularly where space as well as safety are major issues though its attainment might require other approaches for stringent purity criteria.

Nitrogen separation by pressure swing adsorption (PSA) is famous due to its operational flexibility and cost-effectiveness. In this procedure, air gets compressed thereafter passing it through a number of adsorbent beds containing zeolites or activated carbon materials known for selectively absorbing oxygen, water vapour and other contaminants from air while allowing nitrogen to flow as effluent gas. Adsorption occurs under high pressure, usually between 4 and 10 bar. When the adsorbent becomes saturated with oxygen and other pollutants, the pressure decreases, and the adsorbed gases desorb and depart the system. The regeneration process prepares the adsorbent for the following cycle. The PSA process entails the use of adsorption–desorption cycles; this makes it possible to produce nitrogen of high purity continuously. It has some benefits including cheaper operative and maintenance costs, a longer lifespan and fewer required components compared to more conventional nitrogen separation equipment (Guerra et al. 2020; Oxymat: Nitrogen technology pressure swing adsorption (PSA) nitrogen).

9.6 Review of Life Cycle Assessment for E-Fuels

Gasoline e-fuels life cycle assessment (LCA) involves examination of the environmental consequences of each stage of the fuel's production cycle, from extraction of raw materials up to end-use and disposal. It helps in understanding the sustainability as well as environmental impact of e-fuels, especially in the aspect of reducing greenhouse gas (GHG) emissions within transport sector. E-fuels are significant from an environmental point of view, with life cycle analysis (LCA) data showing that GHG reductions of up to 90% compared to fossil fuels can be achieved, if renewable energy is used for the production of e-fuels. Although they are less efficient in well-to-wheel (WTW) terms than direct electrification, e-fuels are considered an important solution for the decarbonization of hard-to-electrify sectors like aviation and shipping. Quantitative analyses reveal that electro-methanol production costs range between 76 and 118 €/MWh, depending on the cost of electrolyzers and capacity factors, with the GHG savings closely related to the carbon intensity of electricity used. The direct air capture of CO_2 is a common step in e-fuel synthesis that increases energy demand, underlining the importance of using low-carbon energy sources to maximize environmental benefits. E-fuels are a type synthetic fuels or electrofuels manufactured by employing electrical energy to transform water and carbon dioxide into hydrocarbons mostly obtained through renewable avenues. LCA for e-fuels commences with energy generation which can be environmentally impactful depending on its source. Renewable energy sources such as wind, solar and hydropower have lower greenhouse gas emissions than electric power based on fossil fuels. The next step is water electrolysis for making hydrogen which then mixes with carbon dioxide to form e-fuel products. This synthesis process usually employs Fischer–tropsch synthesis or methanation that is very intensive in terms of energy and thus necessitates careful examination of energy inputs so as to reduce total greenhouse gas (GHG) emissions.

Additionally, e-fuels transportation and distribution have an environmental impact too. E-fuels may be distributed using the existing infrastructure which might reduce the need to incur new expenses but will still result in emissions from transportation vehicles and probable losses in handling. On the other hand, end-of-life burning of e-fuels in vehicles releases carbon dioxide although this can be matched by carbon dioxide absorbed during production stage hence making the fuel carbon–neutral if all operations are powered by renewable energy. Among the overall LCA there is a need to consider how efficient these engines are as well as how much of emissions they generate while combusting. Comparisons with conventional fuels are part of an LCA. In particular, when renewable power is used, E-fuels often have lower life cycle greenhouse gas emissions compared to fossil fuels. However, production costs and energy requirements remain higher today thereby creating economic or technical obstacles towards their large-scale utilization. The LCA results show that significant reductions of GHG with e-fuels depend on availability of renewable electricity and efficient production technology therefore depending on these factors a lot may change in future and such alternatives will become even more popular than they are now. Policy adjustments and technological advances are required to make e-fuels a viable and sustainable choice in the transportation industry. In conclusion, the LCA of e-fuels demonstrates their potential for considerable GHG reductions, subject to renewable energy consumption and efficient production procedures. This evaluation presents a framework for analysing and mitigating the environmental implications of e-fuels, emphasizing their role in sustainable transportation solutions (Andersson and Börjesson 2021; Ashnani et al. 2015; Liu et al. 2023).

9.7 Literature Review for E-Fuels Production Costs

The literature on the production costs of e-fuels provides useful insights into the economic feasibility and challenges associated with their manufacture.

9.7.1 Overview of Production Costs

1. Investment and Production Cost

Haarlemmer and Boissonnet's research discovered variable production costs for synthetic fuels, including e-fuels. Manufacturing costs are significantly higher, particularly for large-scale operations, and vary widely depending on the technology and feedstock used (Haarlemmer et al. 2014).

2. Cost Predictions:

Costs for producing e-fuels are predicted to fall as technology advances and economies of scale grow. A literature review gives estimates of fuel production costs since 2010, with projections until 2050. These projections account for capital expenditures (CAPEX) for electrolyzers as well as fluctuating power costs (Grahn et al. 2022; Seymour et al. 2024).

9.7.2 *Comparative Analysis*

The production cost of e-fuels and that of conventional fuels differ significantly, as the production costs for e-fuels is more expensive than the conventional fossil fuels with a breakeven oil price estimate ranging from 252–558 $ per barrel due to their high initial production cost related to the water electrolysis, carbon capture and storage and synthesis process provided by the renewable energy. The production costs for fossil fuels ranges between 30–40 $ averagely which makes these fuels comparatively much less costly than the e-fuels. The conventional fuels rely on the existing infrastructure and the e-fuels can leverage the same distribution network with minimum modification.

9.7.2.1 Production Process and Energy Sources

A number of crucial processes are involved in the creation of e-fuels, such as electrolysis, which turns water into hydrogen, and carbon capture, which turns CO_2 into hydrocarbons. The main difference between the e-fuels under comparison is the electrical source that is utilized for the electrolysis process. Renewably produced energy, including hydropower, solar energy, and wind, is essential to reducing the carbon footprint of e-fuels. E-fuels can achieve almost zero greenhouse gas (GHG) emissions when powered by renewable energy. On the other hand, e-fuels' carbon footprint can be greatly increased by utilizing fossil fuels to generate electricity, outweighing any environmental advantages.

9.7.2.2 Efficiency and Carbon Intensity

An additional important comparative factor is the efficiency of e-fuel generation and use. The papers (Andersson and Börjesson 2021; Ashnani et al. 2015; Liu et al. 2023) emphasize that although e-fuels can lower greenhouse gas emissions, production process efficiency as a whole is very important. E-fuels can have their energy efficiency increased and their carbon intensity decreased with the use of technologies like enhanced carbon capture and utilization (CCU) techniques and high-efficiency electrolyzers. Due to the combustion process and related emissions, traditional fossil fuels have a lower overall efficiency and a greater carbon intensity.

9.7.3 Infrastructure and Compatibility

The fact that e-fuels are compatible with current internal combustion engines (ICEs) and infrastructure is one of their main advantages. This dispenses with major adjustments requirements and ensures that the switch from conventional to e-fuels is effortless. However, changing to hydrogen fuel cells or electric cars (EVs) necessitates substantial infrastructure and technological developments leading to a greater upfront cost and a longer transition period.

9.8 Factors Affecting Cost of E-Fuels

There are several key aspects that affect the costs of e-fuels during their life cycle and production. Main factors include renewable energy prices, efficiency of electrolysis process, carbon capture technology and hydrocarbon synthesis (Andersson and Börjesson 2021; Ashnani et al. 2015; Liu et al. 2023; Reddy et al. 2023; Prussi et al. 2022).

1. Renewable Electricity

The main expense in making e-fuel is the electricity necessary for electrolysis. Wind power, solar energy or hydroelectricity have to be utilized since e-fuels must be carbon–neutral. These renewable sources' price and availability have a significant effect on total production costs of e-fuels. Renewable energies would become more widespread as technology advances which will lower cost of electrolysis: this will raise the economic viability of e-fuels.

2. Electrolysis Efficiency

Hydrogen production is dependent on the electrolysis process efficiency, which separates water into hydrogen and oxygen. The technology used for electrolysis such as better designed electrolyzers could lower the amount of energy consumed during the process thereby reducing the cost associated with hydrogen production. Hydrogen generation is a necessary precursor to e-fuels whereby it can be produced at much cheaper costs by use of low power input high efficiency electrolyzers.

3. Carbon Capture Technology

Another crucial issue is whether carbon dioxide should be removed from either the atmosphere or industrial sources in large quantities. There are plans to develop carbon capture and storage (CCS) technologies whose costs vary considerably according to the means of collection and scale of implementation. By raising capture rates through better CCS technologies that reduce energy consumption, we can lower overall e-fuel costs substantially.

4. Hydrocarbon Synthesis

Additions to these expenses occur when synthesizing hydrocarbons from carbon dioxide and hydrogen, for example by using Fischer–tropsch synthesis. The efficiency and size of synthetic reactors as well as the catalysts used are important in determining how cost-effective this conversion can be. These processes can be optimized to reduce costs and increase the economic feasibility of e-fuels.

5. Scale of Production

Economies of scale are very important when it comes to reducing e-fuel costs. Bigger facilities will have lower per-unit expenditures because fixed expenses would be spread over larger outputs, thereby making each unit less costly to produce. The total cost should decline with growing demand for e-fuels alongside increased manufacturing volume.

6. Policy and Regulatory Environment

Government policies and regulations, such as renewable energy subsidies, carbon pricing and low-carbon technology incentives, can have a substantial impact on e-fuel costs. Supportive policies that reduce the cost of renewable energy and encourage the use of CCS can help e-fuels compete with traditional fossil fuels.

9.9 Combustion Characteristics of E-Fuels

The use of e-fuels in conventional engines has the potential to revolutionize climate change mitigation efforts. Unlike gasoline or diesel, these special fuels are created from clean electricity and collected carbon dioxide, resulting in significantly lower emissions—up to 90% less! This is critical since e-fuels operate in existing engines with minor modifications hence, we can continue to use our existing infrastructure while moving to cleaner transportation fuels.

9.9.1 For Solid E-Fuels

The combustion behaviour of metals was originally examined during the development stages of rocket engines. Metal combustion is a unique phenomenon in which metal particles combine with oxygen, forming metal oxides and emitting substantial heat and light. This technique differs fundamentally from the burning of typical fuels such as hydrocarbons because solid oxide products occur at ambient temperature. Magnesium, aluminium and titanium have high combustion temperatures and energy densities, making them appropriate for a variety of high energy applications such as rocket propulsion and pyrotechnics (Bergthorson et al. 2015; Markstein 1963; Grosse and Conway 1958). Particle size, development of oxide layer, and volatility characteristics are physical aspects of metals that could influence combustion process. The flame temperature in metal combustion is often restricted by the boiling point of the metal oxide produced because the evaporation or dissociation of the oxide absorbs additional heat which stops any further rise in temperature (Bergthorson et al. 2015). In addition, fast kinetics or metal–oxygen interactions and refractory nature of metal oxides require special techniques for controlling and supporting combustion making critical capturing of thermal energy generated efficient (Grosse and Conway 1958).

9.9.2 For Liquid E-Fuels

There are promising prospects for sustainable energy solutions from burning liquid e-fuels with possibilities for substantial reduction in GHG emissions and less dependence on fossil fuels. To make Oxy-methylene Ethers (OMEs), which are e-fuels, hydrogen from water electrolysis is usually mixed with carbon dioxide collected from the atmosphere or industrial processes (Boretti 2024). In this way, the process leads to carbon–neutral fuels since CO_2 issued when burning it gets balanced by CO_2 incurred in its production (Boretti 2024). By using present internal combustion engines, e-fuels permit a smooth transition from conventional fuel. The performance of e-fuels has been enhanced by advances in combustion system technologies like direct injection and jet ignition. These systems improve combustion efficiency and reduce emissions, addressing issues such as energy efficiency and scalability (Boretti 2024). For example, in compression ignition engines, OMEs can dramatically reduce nitrogen oxide emissions while increasing efficiency (Novella et al. 2021). The low-sooting properties of e-fuels lead to cleaner combustion, lowering particulate matter and carbon monoxide emissions (Novella et al. 2021). Despite its environmental benefits, the widespread use of e-fuels confronts hurdles, such as high production costs and the need for regulatory support to encourage their use. However, with continued study and technology improvements, e-fuels are predicted to become more cost-effective, serving as an important tool in the global energy shift (Singh et al. 2022). Liquid e-fuels, a sustainable energy source, have the potential to bridge the gap between current fossil fuel reliance and future hydrogen-based energy systems, particularly in areas where electrification is challenging (Boretti 2024; Novella et al. 2021; Singh et al. 2022).

9.9.3 For Gaseous E-Fuels

The combustion of gas e-fuels, such as e-methane, e-hydrogen provides a viable approach to reduce carbon emissions while retaining energy efficiency. E-methane, for example, can be generated from CO_2 and H_2 using the Sabatier process, which normally operates at temperatures ranging from 300 to 400 °C. E-methane combustion produces a high energy production while emitting fewer greenhouse gases than standard fossil fuels (Maas et al. 2016). One of the key benefits of using e-fuels is their compatibility with current internal combustion engines (ICEs) as well as gas turbines that enable gradual shift rather than sudden switch over from fossil fuels to renewable energy sources. When these engines burn e-fuels, they can produce flames of higher temperature particularly when hydrogen is added to the fuel mixture and therefore increase combustion efficiency while reducing CO_2 emissions per unit of fuel (Maas et al. 2016; Huey et al. 2024). Furthermore, no soot or sulphur emission occurs during utilization of e-fuels hence achieving low environmental impact such as soot or sulphur emissions reduction. Moreover, the combustion characteristics and efficiency of e-fuels are enhanced by the possibility to blend several such as ammonia methane and hydrogen. This fuel composition flexibility permits a tailor-made solutions capable of addressing specific energy needs and emission reduction targets. Hence, burning gas-like e-fuels is a feasible and eco-friendly substitute for traditional fuels as part of worldwide endeavours to combat climate change and move toward a low-carbon economy (Maas et al. 2016; Huey et al. 2024).

9.10 Future Scope and Challenges for E-Fuels

Recent technological breakthroughs in the production of e-fuels are now concentrated on combining hydrogen production with carbon capture technologies to make sustainable alternatives to fossil fuels. Innovations such as solid oxide electrolyzers improve the efficiency of hydrogen production by integrating heat and electricity, while advanced carbon capture methods, including direct air capture and post-combustion absorption, enable high purity CO_2 recovery. The combined advancements with cost reduction in electrolysis and carbon capture makes the e-fuels increasingly viable for large-scale deployment, which is supported by numerous global pilot and industrial projects that showcase the potential to significantly reduce emissions while using existing systems. The future of e-fuels lies in significantly contributing to global decarbonization efforts, especially in areas such as aviation, shipping and heavy-duty transportation sectors, where electrification proves difficult due to high energy density requirements. E-fuels are foreseen to be of utmost importance for energy storage by exploiting surplus renewable electricity, thereby ensuring stability of the grid in systems where a large portion of renewables is variable, such as wind and solar. E-fuels are in line with a circular carbon economy since they rely on captured CO_2 for lowering the net greenhouse gas emissions. Their interoperability with current fuel infrastructure enhances their attractiveness as it lowers the cost and complexity of a shift to low-carbon energy systems (Ashnani et al. 2015).

E-fuels pose some challenges that are restraining its widespread adoption since its production is energy-intensive; for instance, the water electrolysis and carbon capture make it a more costly resource compared to the traditional fossil sources. Furthermore, generating e-fuels from renewable sources of energy is less efficient compared to using electricity in its direct applications, like in battery-electric vehicles. Scalability also presents a challenge as high-volume production calls for considerable investments in infrastructures and robust renewable energy supplies. More than that, e-fuels are made from recovered CO_2. Their capacity is determined by the available and efficient carbon capture technology. Their environmental benefits further depend on the low-carbon electricity as using these fuels requires balancing the net-carbon emission. E-fuel acceptance comes with significant social challenges, particularly on labour market conditions in the fossil fuel sector. Transitioning to e-fuels means moving away from the traditional fossil fuel extraction and processing, perhaps even losing jobs in operations such as oil drilling, refining, and coal mining. The change might disproportionately affect regions and communities that currently rely on fossil fuel-related industries for economic stability and work. In addition, the establishment and deployment of e-fuel technologies require different skill sets, such as renewable energy, carbon capture, and electrochemical processes. This will open new doors in renewable energy but calls for an enormous investment in retraining and education for the existing workforce (Mehrpooya et al. 2020; Seymour et al. 2024; Reddy et al. 2023; Prussi et al. 2022).

9.11 Conclusion

The complete scrutiny and evaluation of electrofuels (e-fuels) reveals a lot about their potential as long-term alternatives to traditional fossil fuels in the transport industry. E-fuels, which are produced using renewable energy and captured carbon dioxide have the capacity to reduce greenhouse gas emissions by up to 90% compared with conventional fuels. This shift is necessary for adapting to climate change, particularly in areas such as aviation, heavy-duty transportation, and maritime shipping where electrification is problematic due to high energy consumption and existing battery technology constraints. A good thing with e-fuels is that they can be used together with current internal combustion engines and infrastructure only requiring some few adjustments during installation. There's a more rapid transition from fossil fuels when this is done because it utilizes previous infrastructural investments while reducing environmental hazards greatly. Power-to-Fuel (PtF) technologies that represent the other types of e-fuel production systems comprise numerous synthetic hydrocarbons addressing diverse energy-demanding industries.

Nevertheless, extensive e-fuels utilization brings about worries over scalable production and need for solid infrastructure to support its delivery and consumption. The cost competitiveness of the e-fuels in relation to existing fossil fuels alongside emerging battery technologies remains critical for market uptake. To sum up, research and development (R&D) should be continuous so as to improve efficiency, lower prices and enhance the overall sustainability of processes by which e-fuel is produced. In summary, they can provide a possible way towards a sustainable low-carbon future through giving industrial sectors where electrification is impossible an alternative choice from conventional fossil fuels. For them to be successfully integrated into the global energy landscape, there will be requirement for advances in production technology, sound regulatory frameworks and economic incentives that will lead to their wide spread adoption and use. E-fuels have significant potential in contributing towards global greenhouse gas reduction efforts and climate change mitigation if renewable energy is utilized with current technologies.

List of Abbreviation

AEC	Alkaline Electrolyzer Cell
BECCS	Bioenergy with Carbon Capture and Storage
CAPEX	Capital Expenditure
CCS	Carbon Capture and Storage
CCU	Carbon Capture and Utilization
CO_2	Carbon Dioxide
DAC	Direct Air Capture
e-ammonia	Electro-ammonia
e-fuels	Electrofuels
e-methane	Electro-methane
e-methanol	Electro-methanol
e-methanol	Electro-methanol
EVs	Electric Vehicles
GHG	Greenhouse Gas
H_2	Hydrogen
H_2O	Water
ICEs	Internal Combustion Engines
LCA	Life Cycle Assessment
MJ/kg	Megajoules per Kilogram
NOx	Nitrogen Cxide
OMEs	Oxy-Methylene Ethers
PtF	Power-to-Fuel
PtM	Power-to-Methane
PSA	Pressure Swing Adsorption
R&D	Research and Development
SNG	Synthetic Natural Gas
SOE	Solid Oxide Electrolyzers
SSAS	Solid-State Ammonia Synthesis
SOx	Sulphur Oxide
TW	Terawatt
WTW	Well-to-Wheel

References

Andersson Ö, Börjesson P (2021) The greenhouse gas emissions of an electrified vehicle combined with renewable fuels: Life cycle assessment and policy implications. Appl Energy 289:116621

Arasto A, Onarheim K, Tsupari E, Kärki J (2014) Bio-CCS: feasibility comparison of large scale carbon-negative solutions. Energy Procedia 63:6756–6769

Aricò A, Siracusano S, Briguglio N, Baglio V, Di Blasi A, Antonucci V (2013) Polymer electrolyte membrane water electrolysis: status of technologies and potential applications in combination with renewable power sources. J Appl Electrochem 43:107–118

Ashnani MHM, Miremadi T, Johari A, Danekar A (2015) Environmental impact of alternative fuels and vehicle technologies: a life cycle assessment perspective. Procedia Environ Sci 30:205–210

Bahrami N, Ghandehariun S (2020) Economic assessment of solar-based hydrogen for methanol production. Energy Equip Syst 8(3):263–273

Bergthorson J, Goroshin S, Soo M, Julien P, Palecka J, Frost D et al (2015) Direct combustion of recyclable metal fuels for zero-carbon heat and power. Appl Energy 160:368–382

Boretti A (2024) Advancements in E-Fuel combustion systems for a sustainable energy future. Elsevier, pp 258–266

Brynolf S, Taljegard M, Grahn M, Hansson J (2018) Electrofuels for the transport sector: a review of production costs. Renew Sustain Energy Rev 81:1887–1905

Change IC (2014) Mitigation of climate change. In: Contribution of working group III to the fifth assessment report of the intergovernmental panel on climate change, vol 1454, p 147

Cleary K (2019) Electrification 101. Resources for the Future 5

Daiyan R, MacGill I, Amal R (2020) Opportunities and challenges for renewable power-to-X. ACS Publications

D'Alessandro DM, Smit B, Long JR (2010) Carbon dioxide capture: prospects for new materials. Angew Chem Int Ed 49(35):6058–6082

Elhenawy SEM, Khraisheh M, AlMomani F, Walker G (2020) Metal-organic frameworks as a platform for CO_2 capture and chemical processes: adsorption, membrane separation, catalytic-conversion, and electrochemical reduction of CO_2. Catalysts 10(11):1293

Gibbins J, Chalmers H (2008) Carbon capture and storage. Energy Policy 36(12):4317–4322

Götz M, Lefebvre J, Mörs F, Koch AM, Graf F, Bajohr S et al (2016) Renewable power-to-gas: a technological and economic review. Renew Energy 85:1371–1390

Grahn M, Malmgren E, Korberg AD, Taljegard M, Anderson JE, Brynolf S et al (2022) Review of electrofuel feasibility—cost and environmental impact. Progress in Energy 4(3):032010

Grosse A, Conway J (1958) Combustion of metals in oxygen. Ind Eng Chem 50(4):663–672

Guerra CF, Reyes-Bozo L, Vyhmeister E, Caparrós MJ, Salazar JL, Clemente-Jul C (2020) Technical-economic analysis for a green ammonia production plant in Chile and its subsequent transport to Japan. Renew Energy 157:404–414

Haarlemmer G, Boissonnet G, Peduzzi E, Setier P-A (2014) Investment and production costs of synthetic fuels—a literature survey. Energy 66:667–676

Hawkins AS, McTernan PM, Lian H, Kelly RM, Adams MW (2013) Biological conversion of carbon dioxide and hydrogen into liquid fuels and industrial chemicals. Curr Opin Biotechnol 24(3):376–384

Huey C, Metghalchi H, Levendis Y (2024) E-fuels as reduced carbon emission options. ASME Open J Eng 3

Jimoh TA, Omoarukhe FO, Epelle EI, Okoye PU, Olusola EO, Mukherjee A et al (2023) Introduction to carbon capture by solvent-based technologies. Ref Modul Earth Syst Environ Sci

Julien P, Bergthorson JM (2017) Enabling the metal fuel economy: green recycling of metal fuels. Sustain Energy Fuels 1(3):615–625

Kang S (2021) Innovation outlook: renewable methanol. International Renewable Energy Agency

Levine RB, Pinnarat T, Savage PE (2010) Biodiesel production from wet algal biomass through in situ lipid hydrolysis and supercritical transesterification. Energy Fuels 24(9):5235–5243

Liang ZH, Rongwong W, Liu H, Fu K, Gao H, Cao F et al (2015) Recent progress and new developments in post-combustion carbon-capture technology with amine based solvents. Int J Greenhouse Gas Control 40:26–54
Liu F, Shafique M, Luo X (2023) Literature review on life cycle assessment of transportation alternative fuels. Environ Technol Innov 32:103343
Maas H, Schamel A, Weber C, Kramer U (2016) Review of combustion engine efficiency improvements and the role of e-fuels. Internationaler Motorenkongress: Mit Konferenz Nfz-Motorentechnologie. Springer
Markstein GH (1963) Combustion of metals. AIAA J 1(3):550–562
Mathieu P (2006) The IPCC special report on carbon dioxide capture and storage. ECOS 1–3
Mehrpooya M, Ghorbani B, Bahnamiri FK, Marefati M (2020) Solar fuel production by developing an integrated biodiesel production process and solar thermal energy system. Appl Therm Eng 25(167):114701
Nemmour A, Inayat A, Janajreh I, Ghenai C (2023) Green hydrogen-based E-fuels (E-methane, E-methanol, E-ammonia) to support clean energy transition: a literature review. Int J Hydrogen Energy 48(75):29011–29033
Ni M, Leung MK, Leung DY (2008) Technological development of hydrogen production by solid oxide electrolyzer cell (SOEC). Int J Hydrogen Energy 33(9):2337–2354
Nicita A, Maggio G, Andaloro A, Squadrito G (2020) Green hydrogen as feedstock: Financial analysis of a photovoltaic-powered electrolysis plant. Int J Hydrogen Energy 45(20):11395–11408
Noussan M, Raimondi PP, Scita R, Hafner M (2020) The role of green and blue hydrogen in the energy transition—a technological and geopolitical perspective. Sustainability 13(1):298
Novella R, Bracho G, Gomez-Soriano J, Fernandes CS, Lucchini T (2021) Combustion system optimization for the integration of e-fuels (Oxymethylene Ether) in compression ignition engines. Fuel 305:121580
Oxymat. Nitrogen technology pressure swing adsorption (PSA) nitrogen
Pearson R, Turner J (2012) Renewable fuels-an automotive perspective. In: Comprehensive renewable energy; Biomass and biofuel production, vol 5. Elsevier
Prussi M, Laveneziana L, Testa L, Chiaramonti D (2022) Comparing e-fuels and electrification for decarbonization of heavy-duty transports. Energies 15(21):8075
Reddy VJ, Hariram N, Maity R, Ghazali MF, Kumarasamy S (2023) Sustainable e-fuels: Green hydrogen, methanol and ammonia for carbon-neutral transportation. World Electric Veh J 14(12):349
Rego de Vasconcelos B, Lavoie J-M (2019) Recent advances in power-to-X technology for the production of fuels and chemicals. Front Chem 7:454241
Ridjan I, Mathiesen BV, Connolly D (2016) Terminology used for renewable liquid and gaseous fuels based on the conversion of electricity: a review. J Clean Prod 112:3709–3720
Rivarolo M, Bellotti D, Magistri L, Massardo A (2016) Feasibility study of methanol production from different renewable sources and thermo-economic analysis. Int J Hydrogen Energy 41(4):2105–2116
Rodríguez J, Amores E (2020) CFD modeling and experimental validation of an alkaline water electrolysis cell for hydrogen production. Processes 8(12):1634
Seymour K, Held M, Stolz B, Georges G, Boulouchos K (2024) Future costs of power-to-liquid sustainable aviation fuels produced from hybrid solar PV-wind plants in Europe. Sustain Energy Fuels 8(4):811–825
Singh H, Li C, Cheng P, Wang X, Liu Q (2022) A critical review of technologies, costs, and projects for production of carbon-neutral liquid e-fuels from hydrogen and captured CO_2. Energy Adv 1(9):580–605
Stančin H, Mikulčić H, Wang X, Duić N (2020) A review on alternative fuels in future energy system. Renew Sustain Energy Rev 128:109927
Tarascon J-M, Armand M (2001) Issues and challenges facing rechargeable lithium batteries. Nature 414(6861):359–367

Trost T, Horn S, Jentsch M, Sterner M (2016) Erneuerbares Methan: analyse der CO_2-Potenziale für Power-to-Gas Anlagen in Deutschland. Interdisziplinäre Aspekte der Energiewirtschaft 295–312
Ulleberg Ø (2003) Modeling of advanced alkaline electrolyzers: a system simulation approach. Int J Hydrogen Energy 28(1):21–33
Wilberforce T, Olabi A, Sayed ET, Elsaid K, Abdelkareem MA (2021) Progress in carbon capture technologies. Sci Total Environ 761:143203

Chapter 10
Thermodynamic Properties of an Oval Shape New Flame Casing Implemented/ Deployed Over a Gas Stove Burner: A Simulation Study

Pankaj P. Gohil, Rashmita Patel, Sakshi Sarathe, Sujeet Kesharvani, Gaurav Dwivedi, and Prashant Baredar

Abstract Standard domestic burner of cooking stove efficiency still not achieved on the significant remark as reported. In order to investigate the performance of burner of the domestic gas stove by ensuring the utilizing of the different techniques, one of it to conserve the heat energy. In this paper, an attempt has been made to investigate the thermodynamic properties of performance of new oval-based novel flame shied using commercial simulation code, i.e. Ansys CFX. Based on the simulation approach, the results of maximum temperature have been observed of combustion of flame of 1745 and 1931 K for a domestic burner and oval flame casing burner, respectively. Moreover, the flame velocity of the flame casing burner is obtained more of 0.79 m/s compared to a domestic burner of 0.10 m/s. The overall increase in the thermal efficiency of the Flame casing burner is found to be 22.50%. The overall results seem that performance of the oval shape flame casing better and save a fuel significantly.

Keywords Domestic gas stove burner · Oval shaped flame casing · Thermodynamic parameters · Simulation approach · CFX code

---

P. P. Gohil
Mechanical Engineering Department, Sarvajanik College of Engineering and Technology, Surat, India
e-mail: pankaj.gohil@scet.ac.in

R. Patel
Chemical Engineering Department, Sarvajanik College of Engineering and Technology, Surat, India

S. Sarathe (✉) · S. Kesharvani · G. Dwivedi · P. Baredar
Energy Centre, Maulana Azad National Institute of Technology, Bhopal, India
e-mail: sakshisarathe.96@gmail.com

N. Sharma et al. (eds.), *Sustainable, Scalable and Storable E-Fuels for Decarbonising Transport Sector*, Energy, Environment, and Sustainability,
https://doi.org/10.1007/978-981-96-7388-9_10

10.1 Introduction

The significance of energy in our daily lives stems from the fact that it provides essential human services such as lighting, cooking motive power, space heating and cooling, and water pumping (The World LPG Association 2024). One necessary domestic task that uses energy is cooking. Conventionally, the wood and coal were used as fuels in households for cookstoves, however these fuels can be harmful for the indoor environments. Government and international agencies like United Nations (UN), have encouraged the switch to more environment-friendly sources as fuel for cookstoves like Liquified Petroleum Gas (LPG) in order to reduce the issues of indoor air pollution and health problems related to it.

A burner allows a controlled chemical reaction of fuel and oxidizer (usually oxygen from air) to produce heat (Guo et al. 2024). The heat, infrared radiation, or visible light produced by the flame is commonly used (Bordbar et al. 2022). Bunsen burners powered by natural gas burners are commonly used in gas stoves and cooktops in both residential and commercial settings. A gas stove is one that uses combustible gas, such as natural gas, propane, butane, syngas, liquefied petroleum gas (LPG), or another flammable gas, to fuel it (Gao et al. 2023). Due to its high energy content and minimal particle emissions, LPG is frequently utilized in gas stoves (Islam et al. 2021). LPG use for home purposes has long been advocated alongside a corresponding subsidy (Tran et al. 2022). Dahiya et al. (2016) reviewed various studies on the development of LPG burners and found out that the efficiency does not only depends upon the fuel but also on the burner's construction, materials used and the specifications.

Particular the high demand for LPG in India, there is a need to investigate various modifications to improve the performance of the stove. LPG fuel is used in a wide range of domestic, industrial, commercial, and recreational applications. The type of stove plays an important role in the saving of energy. Ramchandra et al. (2000) conducted an experiment in the village of Uttara Kannada district and concluded that with the use of improved stoves, the fuel saving can be upto, 27–42%. Various studies and experiment have been performed to find the efficiency of various types of stove burners. The Deymi-Dashtebayaz et al. (2023) performed numerical and experimentation of the natural gas stove burner and concluded that the efficiency was found out be between the range 59–65%. Similarly, Islam et al. (2022), conducted a study on natural gas burner and found that the modified burner bottom has 20% more efficiency as compared to traditional one. From the studies it can be seen that there is still scope for energy conservation utilizing different types of burners.

Switching to improved devices that are designed to optimize combustion rate (of fuel, ensuring the presence of sufficient oxygen), radian heat transfer (from fire to vessels, keeping them as close to fire as possible), convection (to pass maximum hot gas over vessels, reducing draught), and conduction could save around 27–42% of energy (Akter Lucky and Hossain 2001). It is revealed from it the energy can be saved by modify the devices that are designed to optimize combustion rate and conduction Heat is concentrated near the vessels by using insulating material for the

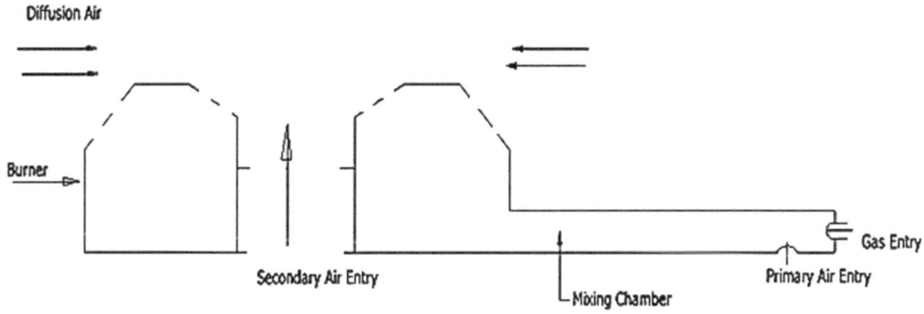

Fig. 10.1 Schematic diagram of a typical conventional LPG cooking stove burner

stove and providing an optimum amount of air to the burner to improve efficiency (Gohil et al. 2022).

Burners are crucial parts of combustion equipment. In order to accomplish full combustion, certain burners have inlets that combine the fuel and air. Acetylene and oxygen are frequently used together for many purposes such as cooking. LPG stoves are straightforward appliances made for burning fuel directly and are commonly used as domestic stove for cooking. It is suited for solid, liquid, and gaseous fuels. For the combustion of gaseous fuels in domestic and industrial heating applications, numerous sorts of burners have been developed. There are two types: (i) total or partially premixed and (ii) nozzle-mix. Domestic gas burners primarily use the Bunsen burner. The schematic diagram of a typical conventional LPG cooking stove burner is shown in Fig. 10.1. A fixed orifice, spud for gas inlet, a controlled shutter for air supply, a venture shaped mixing tube, and a burner head with ports (holes) drilled in it comprise the burner.

With some improvements to existing LPG cooking stoves, a small reduction in consumption per family will result in a massive savings across the country. Therefore, there is a need for research in LPG cooking stoves. A burner is essential in any combustion system for combusting a fuel. Inadequate burner design frequently results in inefficient combustion and excessive pollutant formation.

Jugjai et al. (2001) made a concerted effort to improve cook top burner performance through novel design changes, and concluded that adding a swirling motion to the gas flow in the burner and providing a lighter support for the pot improves efficiency. The majority of cook top burners use flame impingement heat transfer, which means that the distance between the burner and the target (e.g. the bottom of the pot) has a significant impact on heat transfer performance. Hou and Ko (2004) investigated the impact of target height on flame characteristics and overall performance of a domestic gas stove. Various researchers have conducted studies on analytical, numerical, and experimental methods for improving thermal efficiency, reducing pollutant emissions, using various types of fuels, and developing the system to be economically viable. Previously, most studies were conducted using the experimental method to improve efficiency and reduce pollutant emissions (Ko and Lin 2003; Lilley 2012; Tamir et al. 1989; Gohil and Channiwala 2011; Makmool et al. 2007; Stubington et al. 1994; Basu et al. 2008; Ashman et al. 1994). It is the most dependable and practical approach, but it requires more time and money, and it is difficult to analyse the physics of inner flow phenomena. Previously, few studies conducted an insight study of swirl flow in the combustion process. The two optimum angles, one acute of 26° and the other obtuse of 142°, were observed experimentally for synthetic gas, where the thermal efficiency is relatively high and the optimum value is 10° (Gao et al. 2023). Gohil et al. (Dwivedi et al. 2020) investigated the diffuser type flame shield numerically and found that it performs better than conventional gas stove burner. Karyeyen (2018) studied the combustion phenomenon of the new burner.

Various studies have been conducted which Computation Fluid Dynamics (CFD) to analyse the stove burner thermal efficiency and the knowledge of the heat transfer, fluid flow, and phenomenon of combustion. Likewise, Wae-hyee et al. (2021) did the numerical experimentation of various burners using LPG and analysed the heat transfer effect with the help of CFD. Matthujak et al. (2023) used CFD to analyse the performance of modified LPG burner. Moreover, Sangsopha et al. (2024) studied the comparatively analysis of combustion phenomenon of CNG-fuelled and LPG-fuelled ceramic burner using CFD and concluded that the CNG based burner has more temperature as compared to the LPG-based burners.

It is summarized that there is a significant potential for saving energy using the various aspects as reported, and that there is also a need to understand the flow dynamics study of the gas burner flame and enhance the performance of the gas stove burner.

The present study attempted to investigate the thermodynamic parameters of velocity, pressure, and temperature for standard available domestic burner and flame casing burners using the commercially available proven Ansys CFX code. The flame casing was created in the shape of an oval type based on the principle of divergent-convergent section for heat conservation aspect. It is added on top of the standard burner.

10.2 Simulation Methodology

The Computational Fluid Dynamics (CFD) was done using the Ansys CFX software package. The mathematical equations used to carry out the CFD analysis are discussed in this section of the paper is obtainable in the Ansys CFX solver modelling guide (CFD Experts Simulate the Future 2021). The numerical procedure was used in four steps: (1) Construction of Geometry, (2) Generating Mesh, (3) Setting up Boundary Conditions and (4) Simulation.

10.2.1 Geometry of Burners

The first step is to consider the geometry based on the geometrically equal of the available standard domestic burner. Additionally, flame casing burner physical properties are measured and calculated. Based on the data, a complete 3-D model was created in Autodesk Inventor's modeller. The numerical model of the classic burner and flame casing burner is shown in Fig. 10.2.

The inner and outer diameters of the selected conventional burner are 30 mm and 80 mm, respectively, with a total height of 15 mm. The number of holes chosen for the centre side is 30, the outer side is 30, and the upper row and bottom raw are both 60. Figure 10.2a depicts a standard domestic burner. The developed flame casing was placed over the domestic burner (see Fig. 10.2b). The flame casing shape is oval type developed based on the convergent and divergent aspect of the section. The diameters of the bottom and upper sides are same of 80 mm. Based on the existing

Fig. 10.2 Simulation 3-D complete model of domestic burner and flame casing burner

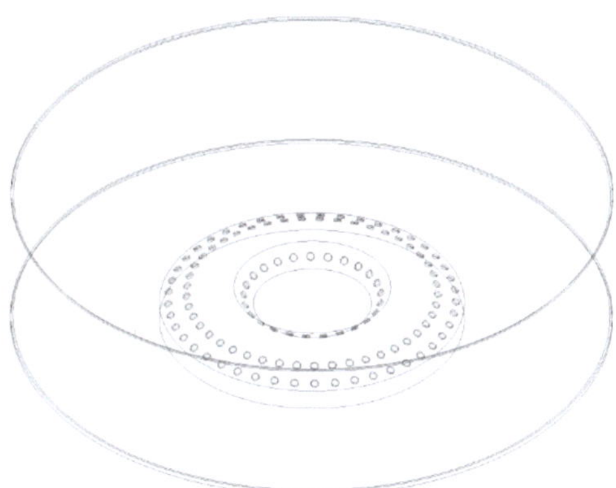

(a) Domestic burner

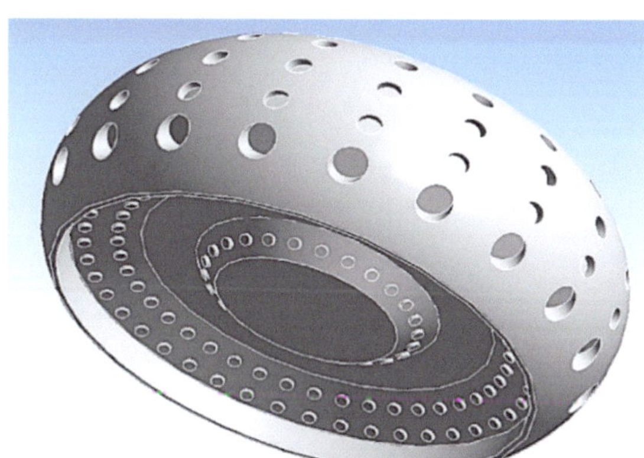

(b) Flame casing burner

structure of a available domestic gas stove, the height of the flame casing is chosen to be 38.5 mm. The total number of holes is chosen from a set of 16 numbers around the perimeter of the flame casing.

10.2.2 Meshing

The second step is to mesh the geometry of the entire part, which is done in Ansys workbench shown in Fig. 10.3. The physical computational domain is meshing with an unstructured grid made up of tetrahedral, prism, and hexahedral elements. Considering all grid quality parameters and maintaining tight grid quality. After running the grid independence test, the meshing value kept between elements is 0.0013m for generating optimized fine mesh for the present study.

Fig. 10.3 Meshing of domestic burner and flame casing burner

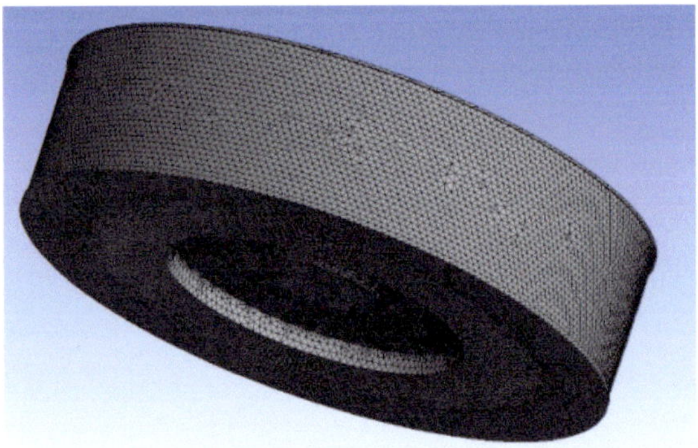

(a) Domestic burner

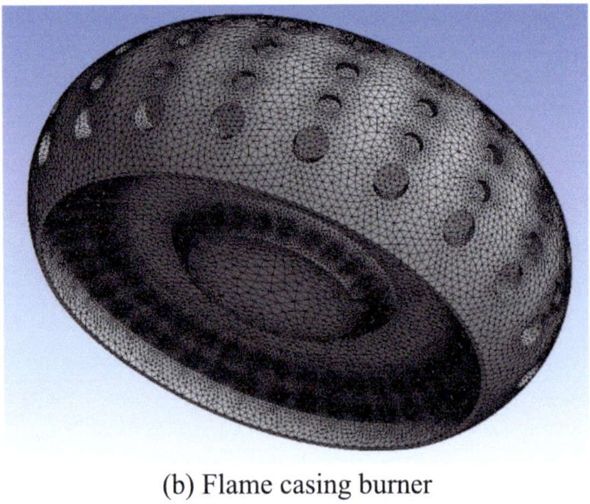

(b) Flame casing burner

10.2.3 Boundary Conditions

Next, the domestic burner and flame casing burner input boundary conditions are calculated using mathematical relationships and experimental data. The mass of air is calculated using the chemical balance equation, which takes into account the LPG fuel's 70% butane and 30% propane composition. The mass fuel is chosen based on the experimental results (Mulugeta Soma et al. 2018). Table 10.1 enlist the various boundary conditions set in the experiment. The fundamental equations for calculating "steady-state incompressible viscosity" of the flow of the fluid with the heat conductivity are the Navier-Strokes equations. While the Navier-Strokes equations are theoretically capable of capturing turbulent flow behaviour, the challenge lies in resolving the wide range of length scales involved, from eddies to extremely small ones. This necessitates extremely fine spatial and temporal discretization in CFD-simulations, often exceeding the current computational resources. To address this, various turbulence models with differing levels of complexity have been developed. RNG k-ε turbulence model is preferred as it is suitable for a broad range, speeds and Reynolds numbers. It was employed to analyse the mixing process of fuel in air in the mixing throat of a domestic cooking burner (Karadimou et al. 2019) (Fig. 10.4).

10.2.4 Simulation Results—Validation

The final step is to simulate the model and obtain and analyse the post-processor results. Validating the model is an essential initial step to confirm its accuracy and the appropriateness of the association setup. The flow pattern was visualized to verify the simulation results. A standard domestic stove burner flame image was taken and matched to a simulation result's streamline. It was observed that a similar trend of fluid streamlines is envisioned for both case scenarios, as shown in Fig. 10.5, The model can be validated with the study conducted by Basu et al. (2008), where similar flame pattern can be seen. The flame consists of two distinct regions: inner flame near each burner port and larger outer flame envelopes surrounding them. The inner-flames are rich premixed flames fuelled by primary air, while the outer flames are non-premixed, resulting from the combustion of unburnt fuel and intermediate species with secondary air. Secondary air is partially drawn from beneath the burner through the central gap and predominantly from the surrounding area above the burner plane. This dual supply of secondary air-radially inward from the exterior and outward through the burner's central opening-creates two non-premixed flame envelopes: an inner one with wine glass-like shape and an outer one with a conical form.

Table 10.1 Detail of input simulation parameters

Analysis type: steady state Fluid type: LPG Reaction mixture: butane air and propane air	
Fluid Model: heat transfer-thermal energy Turbulence—k-epsilon Wall function—scalable	
Inlet boundary	
Mass flow rate of air, g/s	0.587
Mass flow rate of fuel, g/s	0.038
Mass fraction	
C_3H_8	0.7546
C_4H_{10}	0.2453
CO_2	0.1834
H_20	0.0959
N_2	0.7206
Kinematic diffusivity, m^2/s	
C_3H_8	1.2775 E−5
C_4H_{10}	1.038 E−5
CO_2	1.10255E−5
H_20	1.353 E−5
N_2	2.168 E−5
CO	2.073 E−5
O_2	2.221 E−5
Specific heat capacity, J/kgK	1.658
Dynamic viscosity, g/cms	7.4504 E−5
Thermal conductivity, W/mK	0.0138
Diffusivity, m^2/s (LPG-AIR)	1.124 E−5
Outlet boundary	Opening condition: static pressure: 0 Pa and temperature: 273 K

10.3 Result and Discussion

Velocity, pressure, and temperature parameters are evaluated based on the obtained simulation results of domestic and flame casing burner to better comprehend flow phenomena and performance. For the domestic burner and the flame casing burner, simulations were used to produce the streamlines, vectors, and contour of the velocity diagram, which are shown in Figs. 10.6, 10.7, and 10.8 accordingly. It is evident that both instances have uniform streamlines. When a burner has a flame casing, the flow is reduced closer to the central part of the burner than domestic burner (See Fig. 10.6).

It might be because airstrikes are hitting bottom-row shield holes faster and more directly because they are nearer to the burner. Figure 10.7 clearly illustrates flume at

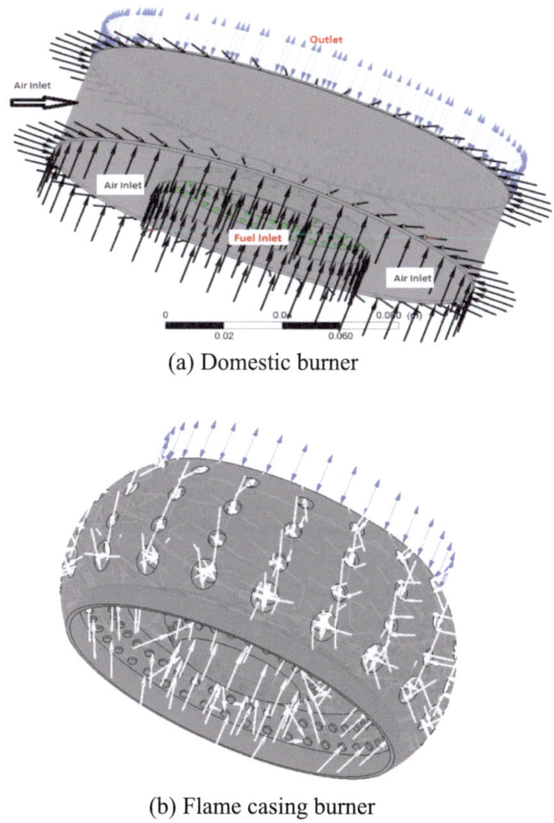

Fig. 10.4 Locations of boundary conditions of the domestic burner and flame casing burner

outer raw of hole is away from the centre diverts to the wall and move to the topside of the burner. Flame bandwidth is found predominant at the top of the burner near the bottom of the vessel.

Moreover, the domestic burner and flame casing burner to have maxima values of velocity of 0.13 and 0.78 m/s, respectively, were achieved. It ought to be noticed that the flame shield burner has found larger velocity region than the domestic burner. It reveals that the bottom side of a vessel gets more heat energy, therefore reduces time food needs to cook. Additionally, Fig. 10.9 shows pressure contours at the mid-plane of burners for both domestic and flame casing burners. In comparison to a standard domestic burner, the flame shield burner's outer edge and centre have more pressure. A maximum pressure zones are observed. The gas is assisted by being pushed toward the vessel's bottom easily.

Finally, the very crucial parameter of temperature is investigated. It demonstrates a direct correlation with the burner's thermal efficiency. The temperature contours at the mid-plane of the burner for the domestic burner and flame casing burner have been observed, and the same is depicted in Fig. 10.10.

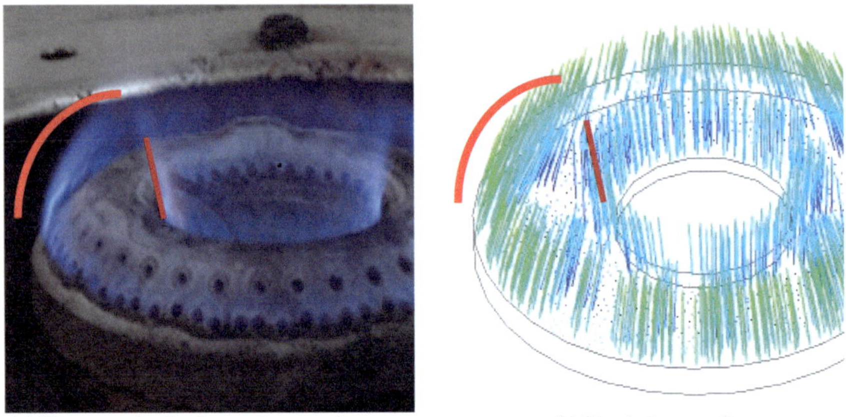

(a) Photograph of a domestic stove burner flame

(b)Simulation result

Fig. 10.5 Comparing of simulation and experiment results

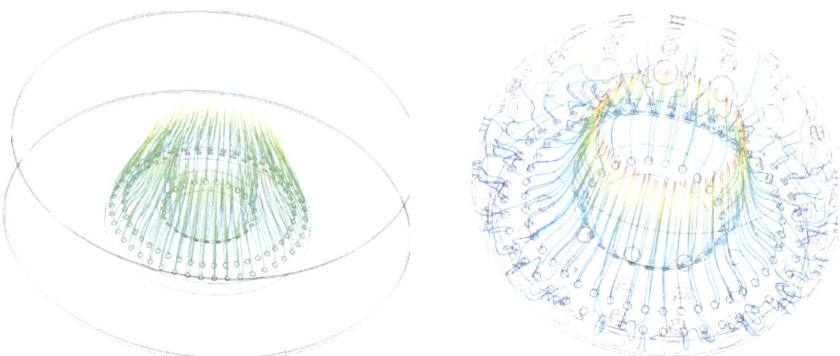

(a) Domestic burner

(b) Flame casing burner

Fig. 10.6 Streamlines of the velocity of flame

(a) Domestic burner

(b) Flame casing burner

Fig. 10.7 Velocity vectors of the flame

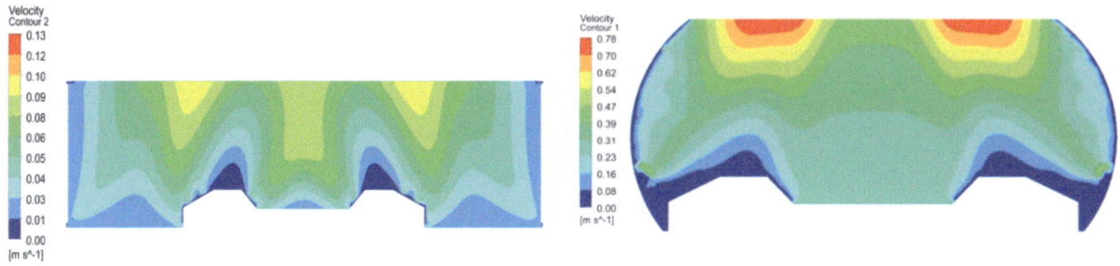

(a) Domestic burner (b) Flame casing burner

Fig. 10.8 Velocity contour of the flame

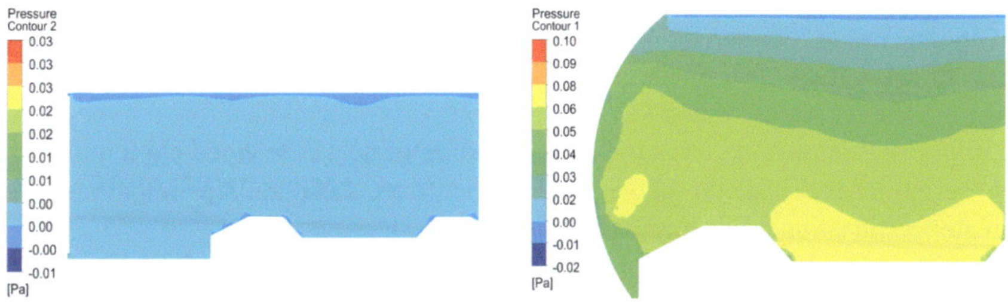

(a) Domestic burner (b) Flame casing burner

Fig. 10.9 Pressure contour for domestic burner and flame casing burner

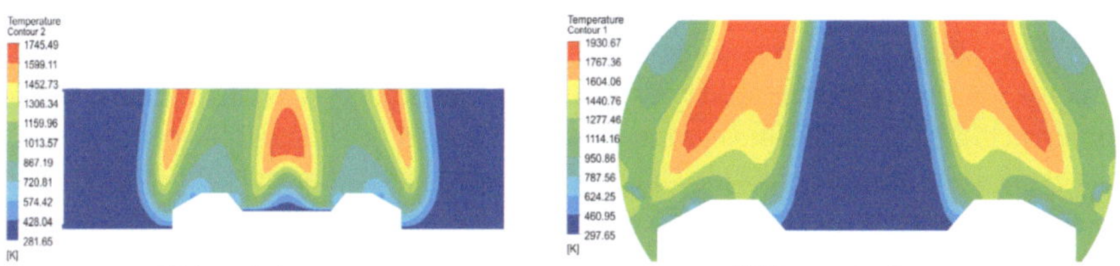

(a) Domestic burner (b) Flame casing burner

Fig. 10.10 Temperature contour for the domestic burner and flame casing burner

It is clearly seen that a flame casing burner causes a greater amount of the temperature-affected area to be visible. The section with a blue line denoted a low heat zone and red colour zone is for high heat magnitude. The lowest temperature has been found at vicinity of centre for flame casing burner. As compared to a domestic burner, the volume of the temperature region is observed to be highest and encompasses a greater area distant from the centre. With a domestic burner and flame casing burner, the maximum temperature is reached at 1745 and 1931 K respectively. With the employment of a flame casing burner, the average temperature increased by 10.66%. This investigation may lead to enhance the thermal efficiency of the burner.

The prepared and summarized overall results, as well as the same outcomes, are shown in Table 10.2. The notable increase in the thermal efficiency is found out to be 22.50% from the overall results.

Table 10.2 Parameters of comparison between the domestic burner and flame casing burner

Parameters	Domestic burner	Flame casing burner
Volume average temperature, K	583	1276
Maximum temperature, K	1745	1931
Average velocity, m/s	0.05	0.61
Maximum velocity, m/s	0.13	0.78
Specific heat capacity at constant pressure, J/kgK	1378	1688
Wall heat transfer coefficient, W/m^{2}K	0.875	3.184

In the case of a flame casing burner, a high specific heat value is noted. The wall heat transfer coefficient inside the burner is achieved higher with employed flame casing. The temperature and velocity both property magnitudes found to be more. It might be demonstrated that the higher amount of heat is retained close to the top of the burner and beneath the vessel. In conclusion, these findings might result in greater combustion efficiency.

10.4 Conclusion

The numerical investigation of standard domestic burner and flame casing burner have been attempted with considering thermodynamic parameters of temperature, velocity, and pressure. Based on the outcomes of the numerical results, a flame shield burner achieves a maximum temperature of 1931 K as compared to a domestic burner of 1745 K. The present study is validated using the previous studies and it shows the similar patterns of flame as the previous studies. The deployment of the flame casing burner led to an increase in temperature of 10.66% on average. Additionally, a flame casing burner achieves a maximum velocity of 0.78 m/s as opposed to the domestic burner of 0.13 m/s. Hence, the overall increase in the thermal efficiency of the flame casing burner is found out to be 22.50%. Compared to the domestic burner, the flame shield burner has a broader velocity region, which should be noted. It demonstrates that a vessel's bottom side receives more heat energy, which shortens the cooking time for food. Additionally, in the case of a flame casing burner, the specific heat and wall heat transfer coefficient were also found to be higher. A comparable line of flow patterns is seen between the results of the experiment and the simulation. The overall findings demonstrate that using a flame casing burner can improve performance while simultaneously preserving energy, making this more commercially viable. The overall results seem that performance of the oval shape flame casing better and save a fuel significantly. The simulation findings it can be seen that the overall thermodynamics properties of the burner have been improved with the introduction of modified burner. The design is commercially acceptable as it shows the similar patterns of flame and more thermal efficiency as compared to the previous studies, which are commercially available for domestic use. The increase in efficiency can reduce the overall fuel consumption and hence can lead to reduction in cost. The durability depends upon the materials used for construction, the burner can be modified with the usage and the highly heat resistant materials can be used with the design in order to increase the durability. The safety can be achieved with the proper handling and functioning of the burner. The burner gives the improved results through which it can be concluded that the burner is highly suitable for domestic use. It would be beneficial to investigate the gas emission parameters for the future study using this same simulation approach.

Acknowledgements Author would like to express his gratitude to Principal, Sarvajanik College of Engineering and Technology for providing the infrastructure and support.

References

Akter Lucky R, Hossain I (2001) Efficiency study of Bangladeshi cookstoves with an emphasis on gas cookstoves. Energy 26:221–237. https://doi.org/10.1016/S0360-5442(00)00066-9

Ashman PJ, Junus R, Stubington JF, Sergeant GD (1994) The effects of load height on the emissions from a natural gas-fired domestic cooktop burner. Combust Sci Technol 103:283–298. https://doi.org/10.1080/00102209408907699

Basu D, Saha R, Ganguly R, Datta A (2008) Performace improvement of LPG cook stoves through burner and nozzle modifications. J Energy Inst 81:218–225. https://doi.org/10.1179/014426008X370951

Bordbar H, Alinejad F, Conley K et al (2022) Flame detection by heat from the infrared spectrum: optimization and sensitivity analysis. Fire Saf J 133:103673. https://doi.org/10.1016/J.FIRESAF.2022.103673

CFD EXPERTS simulate the future (2021)

Dahiya D, Singh Lather R, Bhatia P (2016) Improvement of the domestic LPG cooking stoves: a review. Ind J Sci Technol 9. https://doi.org/10.17485/IJST/2016/V9IS1/105856

Deymi-Dashtebayaz M, Rezapour M, Sheikhani H et al (2023) Numerical and experimental analyses of a novel natural gas cooking burner with the aim of improving energy efficiency and reducing environmental pollution. Energy 263:126020. https://doi.org/10.1016/J.ENERGY.2022.126020

Dwivedi G, Gohil PP, Behura AK (2020) Numerical investigation of thermodynamic parameters for performance evaluation of cooking gas stove burner by appending of flame shield. Mater Today Proc 46:5696–5702. https://doi.org/10.1016/J.MATPR.2020.09.836

Gao W, Hu Y, Yan R et al (2023) Comprehensive review on thermal performance enhancement of domestic gas stoves. ACS Omega 8:26663–26684. https://doi.org/10.1021/ACSOMEGA.3C01628/ASSET/IMAGES/LARGE/AO3C01628_0022.JPEG

Gohil PP, Channiwala SA (2011) Experimental investigation of performance of conventional LPG cooking stove. J Therm Sci Eng 1:25–34

Gohil PP, Dwivedi G, Shukla AK, Verma P (2022) Experimental investigation of heat conservation through novel flame shield arrangement for domestic LPG gas stove. Mater Today Proc 49:223–229. https://doi.org/10.1016/J.MATPR.2021.01.667

Guo JJ, Liu JZ, Zhang T et al (2024) Review on research and development of oxy-coal burner for carbon capture. Sci China Technol Sci 67:647–672. https://doi.org/10.1007/S11431-023-2536-9/METRICS

Hou SS, Ko YC (2004) Effects of heating height on flame appearance, temperature field and efficiency of an impinging laminar jet flame used in domestic gas stoves. Energy Convers Manag 45:1583–1595. https://doi.org/10.1016/J.ENCONMAN.2003.09.016

Islam MM, Wathore R, Zerriffi H et al (2021) In-use emissions from biomass and LPG stoves measured during a large, multi-year cookstove intervention study in rural India. Sci Total Environ 758:143698. https://doi.org/10.1016/J.SCITOTENV.2020.143698

Islam A, Roy H, Rahman MM (2022) Energy efficiency study of household natural gas burner using pot-bottom shield and modified pot arrangement. Energy Rep 8:12871–12885. https://doi.org/10.1016/J.EGYR.2022.09.136

Jugjai S, Tia S, Trewetasksorn W (2001) Thermal efficiency improvement of an LPG gas cooker by a swirling central flame. Int J Energy Res 25:657–674. https://doi.org/10.1002/ER.708

Karadimou DP, Papadopoulos PA, Markatos NC (2019) Mathematical modelling and numerical simulation of two-phase gas-liquid flows in stirred-tank reactors. J King Saud Univ Sci 31:33–41. https://doi.org/10.1016/j.jksus.2017.05.015

Karyeyen S (2018) Combustion characteristics of a non-premixed methane flame in a generated burner under distributed combustion conditions: a numerical study. Fuel 230:163–171. https://doi.org/10.1016/J.FUEL.2018.05.052

Ko YC, Lin TH (2003) Emissions and efficiency of a domestic gas stove burning natural gases with various compositions. Energy Convers Manag 44:3001–3014. https://doi.org/10.1016/S0196-8904(03)00074-8

Lilley DG (2012) Swirl flows in combustion: a review. 15:1063–1078. https://doi.org/10.2514/3.60756

Makmool U, Jugjai S, Tia S et al (2007) Application of particle image velocimetry to analyse thermal performance of LPG cooking burners

Matthujak A, Wichangarm M, Sriveerakul T et al (2023) Numerical and experimental study for a modified LPG cooking burner. J King Saud Univ Sci 35:102752. https://doi.org/10.1016/J.JKSUS.2023.102752

Mulugeta Soma D, Chhabra S, Sehgal SS (2018) CFD analysis of porous medium burner for domestic cooking application. Ind J Sci Technol 11:1–7. https://doi.org/10.17485/ijst/2018/v11i26/130568

Ramachandra TV, Subramanian DK, Joshi NV et al (2000) End use efficiencies in the domestic sector of Uttara Kannada District. Energy Convers Manag 41:833–845. https://doi.org/10.1016/S0196-8904(99)00152-1

Sangsopha A, Matthujak A, Sriveerakul T, Phongthanapanich S (2024). A study on combustion temperature of a rocket-type ceramic burner using CNG as fuel by computational fluid dynamics. https://doi.org/10.3233/ATDE240318

Stubington JF, Beashel G, Murphy T et al (1994) Emissions and efficiency from production cooktop burners firing natural gas. J Inst Energy 67:143–155. https://doi.org/10.1016/0140-6701(95)93648-3

Tamir A, Elperin I, Yotzer S (1989) Performance characteristics of a gas burner with a swirling central flame. Energy 14:373–382. https://doi.org/10.1016/0360-5442(89)90132-1

The World LPG Association. https://www.worldliquidgas.org/wp-content/uploads/2015/10/Guide-to-Good-Industry-Practices-for-LPG-in-Commercial-Kitchens.pdf. Accessed 29 Dec 2024

Tran BL, Chen CC, Tseng WC (2022) Causality between energy consumption and economic growth in the presence of GDP threshold effect: evidence from OECD countries. Energy 251. https://doi.org/10.1016/j.energy.2022.123902

Wae-hayee M, Yeranee K, Suksuwan W, Nuntadusit C (2021) Effect of burner-to-plate distance on heat transfer rate in a domestic stove using LPG. Case Stud Therm Eng 28:101418. https://doi.org/10.1016/J.CSITE.2021.101418

Chapter 11
Leveraging Artificial Intelligence for Sustainable Smart Cities: Insights from Bibliometric Analysis

Geetanjali Kapoor, Meenakshi Singhal, and Pushpendra Chaturvedi

Abstract The adoption of artificial intelligence (AI), information technology and communication has given rise to the concept of "smart cities" with significant implications for several aspects of urban life from infrastructure efficiency to citizen's well-being. Smart cities offer a sustainable solution to urban challenges and promote citizens' social and economic growth. In order to make inhabitants' lives comfortable in diverse sectors, integrating artificial intelligence in sustainable smart cities strive to offer them a smart and conducive environment. There is a need to address the opportunities, identifying key insights, emerging trends and future research directions in this crucial field. This data extraction presents a comprehensive bibliometric analysis of existing research on the intersection of artificial intelligence and sustainable smart cities. This article provides an overview of the application of AI and trial systems that demonstrate how artificial intelligence may play a vital tool in achieving sustainable management in smart cities and further identifies the research hotspots for the implementation feasibility. This work extracts systematic literature review of citation network analysis from the Web of Science database using several performance measures. With the aid of VOS viewer, the co-authorship between authors and nations is evaluated. The findings reveal that AI in context of smart cities is an emerging field and proposes future research scopes based on study mapping and outlines the research expansion of the significance of artificial intelligence (AI) in Sustainable Smart Cities.

G. Kapoor (✉)
Chandigarh University, Mohali, Punjab, India
e-mail: geetanjali.e9170@cumail.in

M. Singhal
Guru Nanak Dev University, Amritsar, Punjab, India
e-mail: meenakshi.arch@gndu.ac.in

P. Chaturvedi
Malaviya National Institute of Technology, Jaipur, India
e-mail: pkchaturvedi8561@gmail.com

N. Sharma et al. (eds.), *Sustainable, Scalable and Storable E-Fuels for Decarbonising Transport Sector*, Energy, Environment, and Sustainability,
https://doi.org/10.1007/978-981-96-7388-9_11

Keywords Artificial intelligence · Sustainable smart cities · Bibliometric analysis · VOS viewer · Future research

Nomenclature

AI	Artificial intelligence
ICT	Information and communication technology
IoT	Internet of things
UC	Unit commitment
MC	Machine learning
DL	Deep learning
NILM	Non-intrusive load monitoring

11.1 Introduction

11.1.1 Background

The rapid urbanization and fast expanding population in the twenty-first century has made it extremely challenging to run the cities effectively and efficiently. Cities are continuously facing challenges in providing top-notch services, ensuring the sustainable management of resources and improving the overall quality of life of citizens. In the said context, the notion of "smart cities" has emerged as a viable approach to deal with the challenges through the integration of cutting-edge information and communication technologies such as artificial intelligence and the Internet of things. By leveraging a wide range of technologies smart cities manage to optimize urban infrastructure, resources and services efficiently. In cities like China, AI-powered technologies are being used to anticipate and avoid air pollution, reduce emissions by 20% as well as control; traffic signals more effectively. Smart cities can enhance the quality and effectiveness of services, encourage sustainable economic growth and eventually improve the lives of the residents by using AI and other intelligent systems. Nevertheless, there are certain obstacles in implementing AI in smart cities. Despite the growing significance of smart city initiatives on the legislative agenda, few cities have genuinely become "smart," with the majority still in early phases of implementation. In order to overcome these challenges and realizing the full potential of AI-driven smart cities, a comprehensive knowledge of the relevant technologies, their applications and potential effects on the overall environment and governance is essential. This literature review through bibliometric analysis will pave a way to identify the potential hotspots and future research directions in the field of sustainable smart cities and artificial intelligence.

11.1.2 Rationale of the Research

The study supports leveraging artificial intelligence for sustainable urban development through bibliometric analysis to explore emerging research trends. The study further seeks to explore the synergies between AI and Sustainable cities as a part of efficient planning to tackle operational challenges in the cities. A bibliometric analysis is a research methodology for systematically investigating the origin, scope, and impact of research in the interdisciplinary field. A wide range of academic publications is analyzed to identify the significant areas to be explored further. The rationale for this research lies in the following aspects:

- Addressing Knowledge Gaps: A comprehensive overview of the larger research landscape is absent that highlights the role of AI in urban concerns specifically. The study bridges the gap by mapping the latest trends and kn knowledge clusters in the related field.
- Policy Implications: The findings through bibliometric analysis can assist the policy makers and urban planners in leveraging AI technology for sustainable urban growth.
- Future Research Directions: The research offers a roadmap to identify emerging studies and highlights he underexplored topics and further promotes the interdisciplinary collaborations between city development experts, urban planners and AI researchers.
- Social Relevance: To accomplish the sustainability targets such as Sustainable Development Goals (SDGs) of United Nations, the research underlines the crucial role of artificial intelligence (AI) in addressing urban issues in alignment to global objectives.

This work will add to sustainable urban development research and highlight AI's potential to transform cities into resilient, inclusive, and environmentally friendly places.

11.1.3 Literature Review

Artificial intelligence (AI) has affected almost every aspect of life, including marketing, banking, and finance, as well as industries like agriculture, healthcare, security, robotics, and transportation. AI now makes it possible for solutions to the sustainability that human civilization has been attempting to achieve since its inception (Yigitcanlar et al. 2021). Current advances in AI have pushed the so-called smart city to transform into an independent, self-healing system. Largely uncharted urban beast AI is taking control over urban services and urban governance entities in this newly arising branch of smart urbanism, allowing the city to run on its own (Cugurullo et al. 2021). Numerous technological breakthroughs in new innovations, such as the Internet of things (IoT), AI, and machine learning (ML), are employed to create the ecosystem needed to execute apps and create campuses that are efficient, smart, and sustainable. These enormous campuses are scale models of modern cities (Agarwal et al. 2020). The edge AI provides in the case of energy-based services for a smart city may possess the smart metering and non-intrusive load monitoring (NILM) operating procedures. These applications will help in energy use and sustainability of urban spaces (Chui et al. 2018). Growing cities encounter a variety of difficulties, related to security, transport, sustainability, resource allocation, and public health. Residents can participate in the urban digital environment through software applications, recognized by policymakers and researchers as an opportunity (Cugurullo et al. 2021). In one of the studies, by examining the survey results on AI-run transport systems, the study for smart cities looks forward to a three-fold review of its public acceptance and sustainability approach (Al-Marghilani 2022). Technologies like artificial intelligence, the Internet of things, and many others have recently advanced, enabling the creation of smart cities and resonating with sustainability concepts like the use of clean and renewable energy while taking into account the economic, educational, health, social, and environmental aspects of the city (Albrecht et al. 2015). Although AI has made our lives easier by providing deeper access to sectors like e-commerce and banking, on the other hand, it has increased the proximity of cyberbullying. Artificial intelligence models like machine learning and deep learning (DL) can be employed for the detection of cyberbullying (Chui et al. 2017). The smart city implies a vision for sustainable development with the help of artificial intelligence (AI), the Internet of things (IoT), big data, and information and communication technology (ICT). These technologies can also help in the health care sector in diagnostics like cardiovascular, diabetes, Alzheimer, and other forms of dementia (Allam and Dhunny 2019). The studies aim to explore the phenomenon of platformization and refers to the development of the parallel virtual world known as the metaverse. The infinite possibilities that this reconstructed reality would provide could take the urbanization and sustainability goals either way (Anthony 2021). The viability of human existence is gravely threatened, and this risk affects our economy, society, and environment. Smart cities and societies have the potential to bring people and countries together in the pursuit of sustainability because they require interaction with our environments, assessment of those environments, and the making of sustainable decisions guided by the principles of the triple bottom line. Unhealthy healthcare systems have an impact on people, society, the environment, and economies. In order to automatically find healthcare services that may be produced or co-created by multiple stakeholders using social media analysis, this studies offer a data-driven AI-based technique called MUSAWAH (Alahmari et al. 2022). Policymakers, urban managers, and planners can gather, monitor, analyze, and assess the circularity of environmental, social, and economic resources to increase their efficacy and quality with the help of digital technologies that are involved in urban metabolism for efficiency (D'Amico et al. 2021). The utilization of renewable energy sources (RES) and unit commitment (UC) is the most successful artificial intelligence-based approach for energy sustainability. Thus, a different approach to solving the UC issue is presented. The suggested strategy includes a straightforward procedure to produce the well-liked solutions in a reasonable amount of time by building a fundamental model of the schedule of the condition of energy units. It is evident that distinct consumer units with an operational focus evaluate their hourly performance metrics economically in order to reduce costs. This is accomplished by employing an artificial intelligence algorithm to optimize the specialized cost function, which is mimicked by announcing. This optimization is achieved through the application of an advanced artificial intelligence algorithm, which dynamically adjusts to optimize the specialized cost function. The algorithm operates by simulating real-world conditions, adapting to variations in energy demand and supply, and intelligently announcing optimized schedules. This process not only enhances cost-efficiency but also contributes to the sustainable management of energy resources (Garlik 2022). A popular and precise technique for looking through and evaluating vast amounts of scientific data is bibliometric analysis. It allows us to examine the intricate details of a particular topic's advancement while illuminating new developments within that field (Donthu et al. 2021). Although previous research has provided detailed guidelines for conducting a bibliometric research, further information is needed to enhance the effectiveness and design of this bibliometric approach, underscoring the need for more investigation (Öztürk et al. 2024).

11.1.4 Objectives

The primary goal of the research is to investigate the published documents in the last 11 years for a detailed survey based on bibliometric analysis of the emerging drift of integration of artificial intelligence in the progression of smart cities and attempting to make them sustainable. The substantial contributions of the most relevant authors are as follows:

- Literature analysis of applications of urban artificial intelligence in diverse systems of smart cities.
- Technological advancements for transforming the dynamics (mechanics and physicality) of initiatives of the smart city.
- Contribution of the Internet of things (IoT) in the exchange of information and communications in smart cities.
- The research trends analysis on the basis of sources, authors, and keywords.
- Survey of literature of 20 most relevant documents cited at the top for AI and Smart city.
- Analysis of top affiliations and journals contributing to research.
- Main obstructions and gaps for artificial intelligence integration in sustainable smart cities.

11.1.5 Research Questions

The study is an outcome of collaborative contributions to the literature on the applications and integration of artificial intelligence in making cities smart and sustainable. A thorough bibliometric analysis of papers published on artificial intelligence (AI) by various authors, nations, journals, and organizations during the past 11 years is used to analyze the data. The majority of the most cited papers about 15–20 have been examined. The following are the two main and one supporting research questions:

1. Research Question: What are the most recent uses of artificial intelligence in the construction of smart cities around the world?
2. Research Question: How does artificial intelligence help in making cities smart and sustainable?
3. Research question: What are the underrepresented areas in the overall implementation?

11.1.6 Research Novelty

The study is innovative in its approach to analyze the intersection of artificial intelligence (AI) and sustainable urban development systematically through bibliometric lens. In contrast to traditional reviews that mainly concentrate on qualitative insights, this study employs the quantitative bibliometric tools to uncover the knowledge gaps, hidden patterns and emerging trends in the existing body of literature.

The study offers a distinctive, data-driven perspective on cutting-edge technologies that are reshaping the urban planning by fusing AI applications with urban sustainability frameworks. This groundbreaking work establishes the study as an essential tool for scholars, professionals, and decision-makers.

11.2 Methods

The search was done for the topic with the keywords" Artificial Intelligence" and "Sustainable" and "Smart City" in the database of the web of science from 2011–2023 before the launch of the smart city Mission in India. VOS viewer and bibliometric analysis offers a systematic method for identifying the research gaps and trends in the application of artificial intelligence (AI) for sustainable smart cities. This approach helps track the publication tends, citation impact, and emerging themes wherein AI can improve sustainability. The motive of the research is to inspect the publications demonstrating the relevance of artificial intelligence (AI) in Smart city development through a sustainable approach. The related journals, authors' works, documents, book chapters and other similar articles. This methodology supports the subject

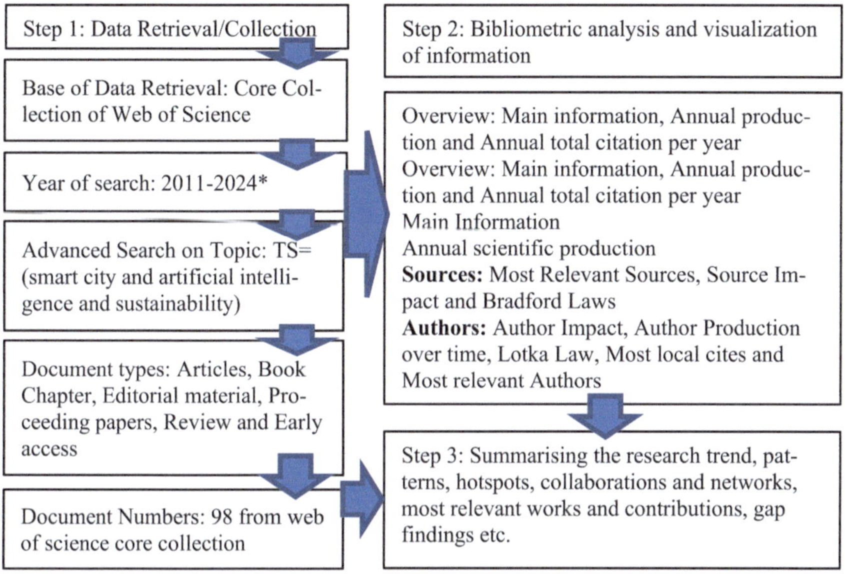

Source: Authors

Fig. 11.1 Bibliometric analysis steps on Topic Search (TS) = ("smart city" and "artificial intelligence" and "sustainability") on WoS database

by providing data-driven insights that might leverage to promote urban sustainability, pinpointing unexplored areas for future research. The search was conducted on January 16, 2023, and is shown in Fig. 11.1.

11.3 Results and Discussions

11.3.1 Publication and the Yearly Production of Documents

The yearly production and publication along with the growth trends help to uncover and indicate the burgeoning research areas. The graph as an output shows how frequently the articles are being cited and published based on their quality. Figure 11.2 Describes the timeline of the articles cited and published on TS = (smart city and artificial intelligence and sustainability). In addition to this Table 11.1 lists out the ten most productive sources mentioning rank, source and number of articles and Table 11.2 lists top 10 most productive organizations and countries among 98 publications.

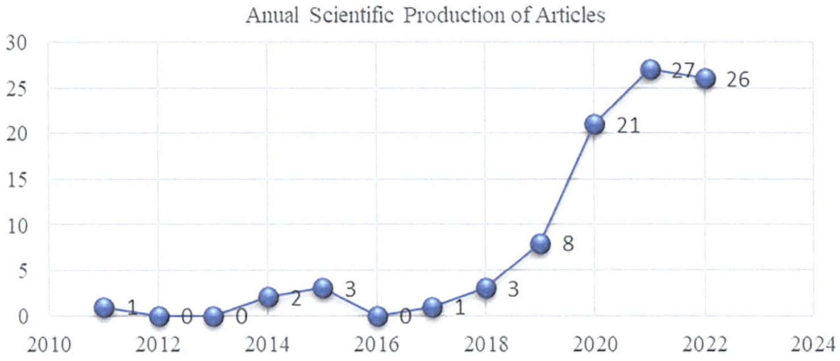

Fig. 11.2 Year-wise production: number of publications from 2011–2023* on TS = (smart city and artificial intelligence and sustainability)

Table 11.1 List of 10 most productive sources, and authors among 98 publications between 2011–2023, in chronological order

Rank	Source titles	Articles	Authors	Articles
1	Sustainability	22	Yigitcanlar T	6
2	IEEE Access	5	Allam Z	3
3	Sustainable Cities and Society	4	Cugurullo F	3
4	Energies	3	Agarwal P	2
5	Smart Cities	3	Bibri Se	2
6	AI and Society	2	Chui Kt	2
7	Journal of Cleaner Production	2	D'amico G	2
8	Sensors	2	Garlik B	2
9	2014 12th IEEE International Conference on Industrial Informatics (India)	1	Ioppolo G	2
10	2019 International Conference on Computing, Networking, And Communications (Icnc)	1	Mehmood R	2

Table 11.2 List of 10 most productive organizations and countries among 98 publications between 2011–2023, in chronological order

Rank	Affiliation	Articles	Countries	Articles
1	Queensland Univ Technol	7	China	36
2	Univ Calif Berkeley	6	Usa	29
3	Deakin Univ	4	India	28
4	Hiroshima Univ	4	Australia	25
5	King Abdulaziz Univ	4	Uk	23
6	Kth Royal Inst Technol	4	Spain	20
7	Norwegian Univ Sci And Technol	4	Italy	18
8	Rajiv Gandhi Inst Petr Technol	4	France	12
9	Univ Politecn Madrid	4	Saudi Arabia	12
10	Carnegie Mellon Univ	3	Canada	7

11.3.2 Keyword Analysis

Keywords were extracted from the open-access platform VOSviewer that reflected the collection of nouns or words from the core publications of WOS database. It depicts the frequency of the articles with the keywords has been cited in different articles along with the scientific analysis of impacts. One of the metrics used to evaluate the potential of research and publications in sustainability and other well-known scientific journals is the analysis of keywords. The bibliometric analysis shows that there was total of 656 keywords and with a frequency of "3" times a keyword appears, 72 meet the threshold for network visualization as shown in Fig. 11.3. Different-sized circles depict the keyword occurrences. The larger the size the more frequently the keyword is co-selected in the articles of study area. The keyword "Artificial Intelligence", "Sustainability", "Smart Cities", and "Smart City" represent the strongest strength. The topic similarity and relative strength are represented by the distance between the keywords.

11.3.3 Keyword-Tree Map Analysis

The top 20 keywords in the word tree, which are the discovered hot-spot keywords given in Fig. 11.4, reflect the rapidly expanding subjects in the field of smart city research. The most occurring keywords are "Big Data" represented in dark blue color in the tree with 6% occurrence, "Cities" and "City" represented in orange and red in the given tree has an occurring rate of 5%. Whereas keywords like "smart cities" and "Sustainability" are represented by cyan and navy blue in the word tree only occurring with a percentage of 2%. The word tree establishes the fact that there is a gap between smart cities and artificial intelligence. There is an urgent need to have a trend in smart cities and sustainability and artificial intelligence because of the residents' needs for daily habitation.

The top 10 keywords in the word tree, which are the discovered hot-spot keywords given Table 11.3, reflect the rapidly expanding subjects in the field of smart city research. The most occurring keywords are "Big Data" represented in dark blue color in the tree with 6% occurrence, "Cities" and "City" represented in orange and red in the given tree has an occurring rate of 5%. Whereas keywords like "smart cities" and "Sustainability" are represented by cyan and navy blue in the word tree only occurring with a percentage of 2%. The word tree establishes the fact that there is a gap between smart cities and artificial intelligence. There is an urgent need to have a trend in smart cities and sustainability and artificial intelligence because of the residents' needs for daily habitation.

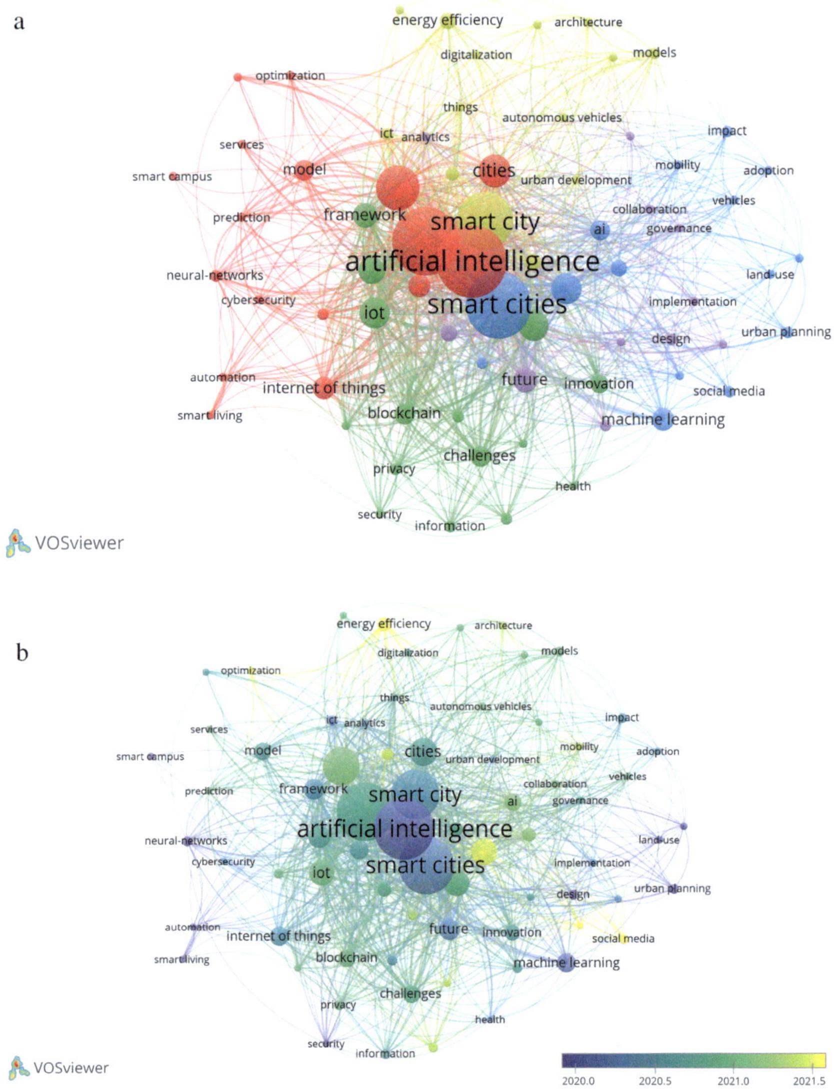

Fig. 11.3 Co-keyword network visualization using VOS viewer on smart cities research. *Note* **a** co-keyword network visualization was based on occurrences (frequencies shown as colored legend); **b** co-keyword overlay visualization was based on the occurrences and average publication per year scores (global scenario year of publications shown in colored legend)

Fig. 11.4 Keywords tree map on TS = (smart city and artificial intelligence and sustainability) research

Table 11.3 Correlation strength of the top 10 occurrence keywords

Rank	Keyword	Occurrences	Total link strength	Cluster
1	Artificial intelligence	36	167	1
2	Sustainability	34	174	1
3	Smart cities	33	176	3
4	Smart city	28	164	4
5	Big data	20	151	4
6	Cities	14	67	1
7	City	13	110	2
8	Artificial intelligence	13	91	3
9	IoT	13	88	2
10	Internet	11	104	2

11.3.4 Co-authorship Visualization Analysis

The co-author network visualization network analyzed the collaboration and contribution pattern of author and further the organizations and countries that are publishing on the same search topic. On the basis of 98 publications, 343 different authors contributed in publishing and in the network study it was identified that with a minimum one number of document and with minimum 5 citations per author only 181 met the thresholds. The largest collection of linked things consists of 23 items and the network visualization shows the strength in Fig. 11.5.

All of the top 10 writers with the strongest co-authorship ties were published between 2011 and 2023. The Table 11.4 shows that "Yigitcanlar T" has published 6 articles in the year 2021 with a total count per year of 43 and author "Allam Z" has published 3 articles in the year 2022 with a total count per year to be 50.6. It can be observed from the table that very less articles have been published on smart cities and artificial intelligence and sustainability as in Table 11.4.

11.3.5 Analysis Based on the Cooperation of Countries and Regions

Analysis of visualization of co-author of different countries/regions. Table 11.5 shows the bibliographic extracted data from the Web of Science (WoS) core collection on the basis of the network visualization of cooperation and co-authorship on VOS viewer. The process shows that out of 51 countries keeping minimum number of documents of a country with minimum 3 nos citations, 18 meet the thresholds.

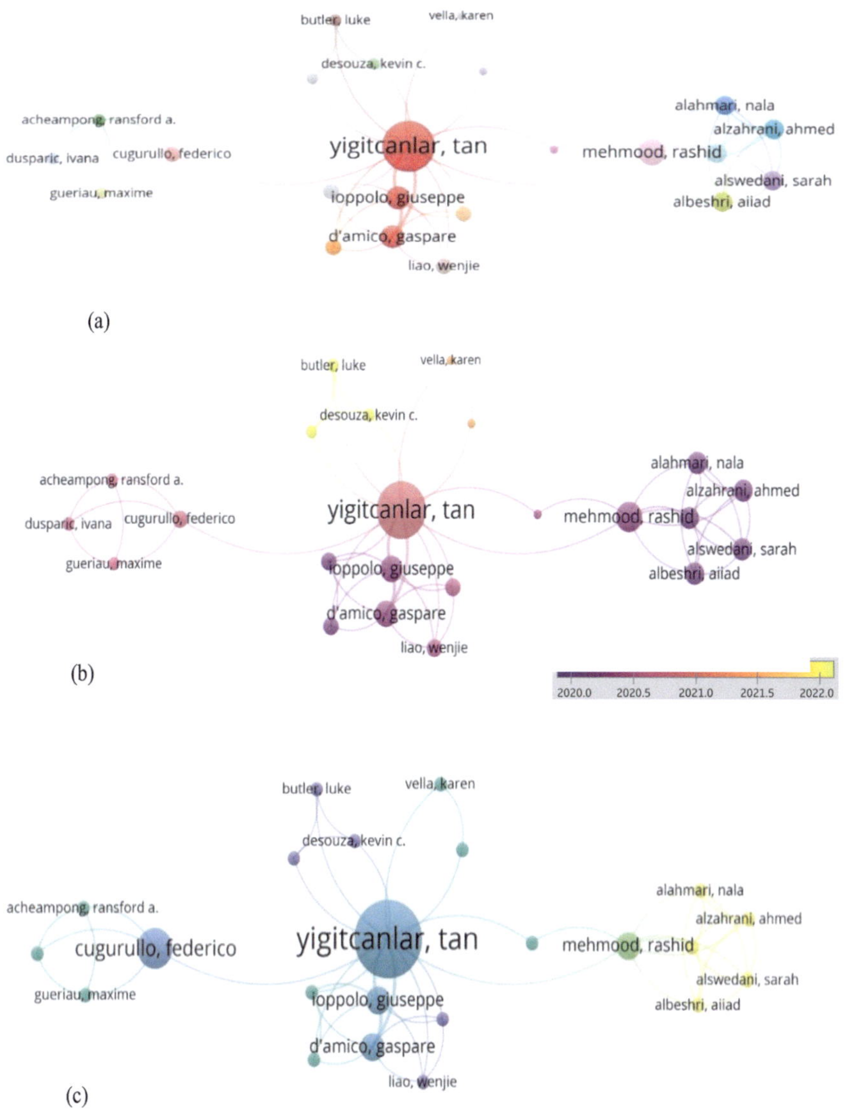

Fig. 11.5 Co-authorship network on TS = (smart city and artificial intelligence and sustainability) research. *Note* **a** author link strengths shown as articles fractionalized, **b** VOSviewer overlay visualization showing total citation per year on an average basis, **c** VOSviewer overlay visualization showing link strengths with citation scores

Table 11.4 The top 10 strong co-authorship links of authors published recently in 2020–2022 in the domain of smart city and artificial intelligence and sustainability research and growing exponentially

Authors	Articles	Articles fractionalized	Year	Tc	Freq	Tcpy
Yigitcanlar T	6	1.82	2021	172	3	43.000
Allam Z	3	0.95	2022	253	2	50.600
Cugurullo F	3	1.75	2021	91	2	27.750
Agarwal P	2	0.67	2020	12	2	3.000
Bibri Se	2	0.45	2022	10	2	5.000
Chui Kt	2	0.5	2018	51	1	12.500
D'amico G	2	0.4	2021	26	1	6.500
Garlik B	2	2	2022	0	2	0.000
Ioppolo G	2	0.4	2021	26	1	6.500
Mehmood R	2	0.5	2022	24	1	8.000
Yigitcanlar T	6	1.82	2021	172	3	43.000

Table 11.5 Prominent cooperation network of countries/regions along with characteristics

Country	Documents	Citations	Total link strength	Freq
Australia	16	569	30	25
Peoples R China	15	251	21	36
Spain	14	249	13	20
Saudi Arabia	10	200	13	12
Italy	11	148	23	18
Ireland	4	130	9	5
England	15	110	16	23
Greece	4	99	3	5
USA	9	69	15	29
India	10	68	17	28
Australia	16	569	30	25

11.3.6 Analysis Based on the Cooperation of Countries and Regions

Co-author visualization analysis of countries/regions: Fig. 11.6 shows the bibliographic extracted data from the Web of Science core collection on the basis of the network visualization of cooperation and co-authorship on VOSviewer. The process shows that out of 51 countries keeping minimum number of documents of a country with minimum 3 nos. citations, 18 meet the thresholds. Furthermore, Table 11.5 represents prominent cooperation network of countries/regions along with characteristics

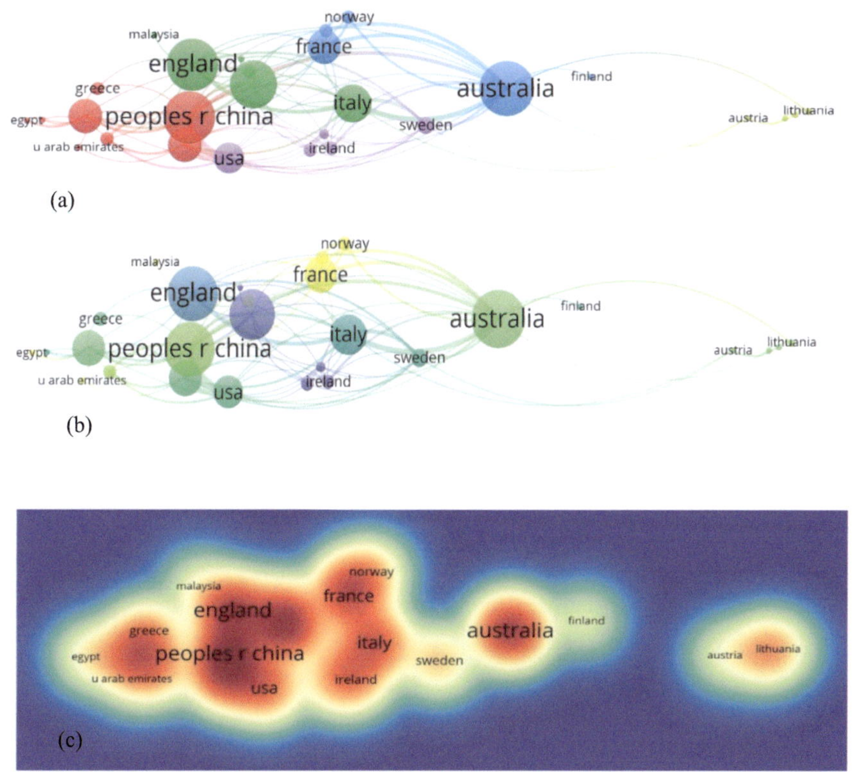

Fig. 11.6 Map of countries co-author visualization with document weights with network visualization map and density visualization map (**a**–**c**)

such as country name, number of documents, total citations, total link strength, and frequency (as shown in previous section).

11.3.7 Comparative Observations of Smart City Search World-Wide and in Indian Context in Year 2023

While analyzing the keywords trends on the world level, it can be seen that the keyword "smart city" has occurred 404 times with total link strength as 1317 in the year 2023 with "smart cities" and "Cities" occurring 272 and 220 times respectively with total link strength as 933 and 838 in the order of occurrence. The top 10 keywords list also includes words like "Internet", "Internet of things", "challenges", "governance", "big data", and "IoT". The analysis shows the lack of keywords like "artificial intelligence", "sustainability", "industry 4.0", etc. and is evident in Fig. 11.7a.

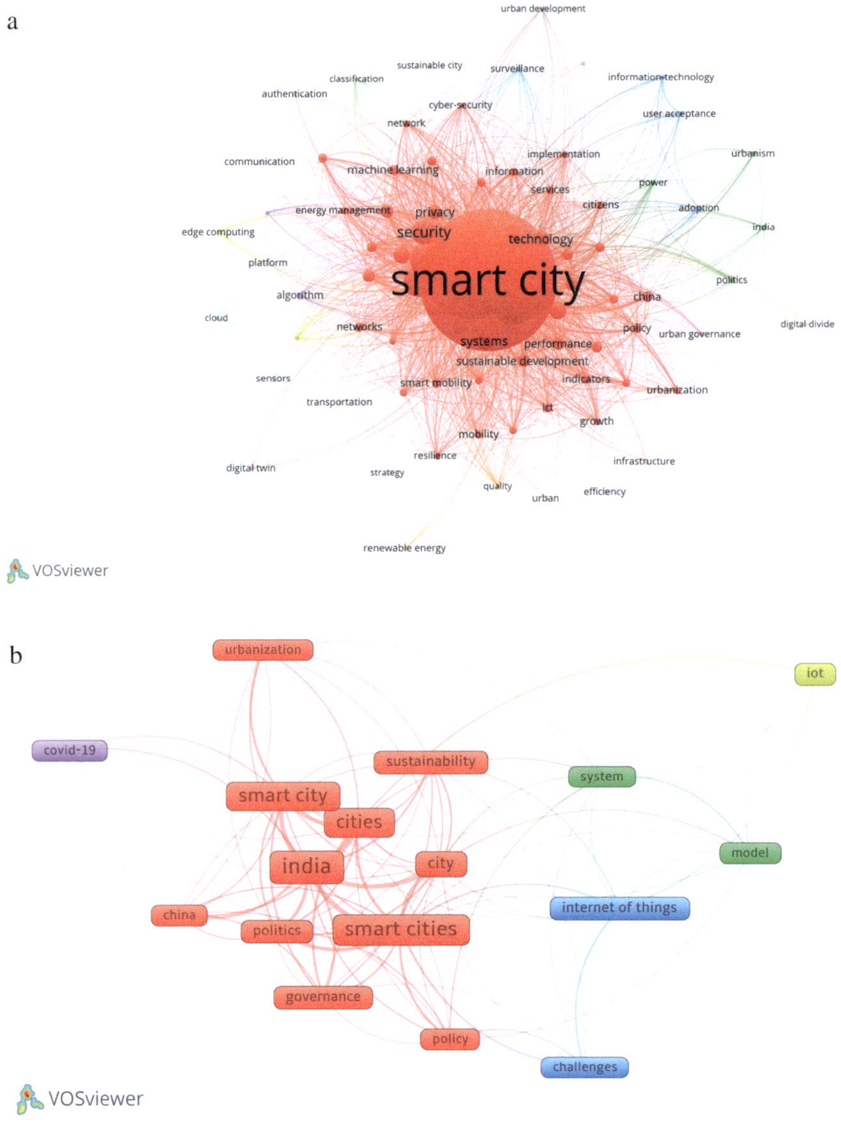

Fig. 11.7 **a** Shows the keywords observations of smart city search in context to world-wide in the year 2022 **b** shows the keyword observation in context to India in the year 2022

In the Indian context, the most occurred keywords are "India", "smart cities", and "smart city" with occurrences as 25, 23, and 19, respectively, and is represented in Fig. 11.7b. Other top 10 keywords are "governance", "Internet of things", "IoT", "politics", and "sustainability". In India, it can be seen that the smart city is

Table 11.6 (a) Top 10 keywords occurrence related to the world search in the year 2022 (b) Top 10 keyword occurrences related to India in 2022

Rank order	Keyword (World-wide)	Occurrences	Total link strength	Keyword (India Only)	Occurrences	Total link strength
1	Smart City	404	1317	India	25	64
2	Smart Cities	272	933	Smart Cities	23	58
3	Cities	220	838	Smart City	19	40
4	City	138	608	Cities	18	46
5	Internet	120	665	City	12	34
6	Internet of Things	80	382	Governance	8	23
7	Challenges	79	480	Internet of Things	8	18
8	Governance	79	363	IoT	8	13
9	Big Data	77	450	Politics	8	28
10	IoT	77	384	Sustainability	8	19

very loosely interacting with words like IoT, artificial intelligence, Internet 4, and sustainability as shown in Table 11.6.

11.3.8 Affiliations Co-operations on TS = (Smart City and Artificial Intelligence and Sustainability) Research

In Fig. 11.8, the top-ranking ten affiliations/organizations have been listed. The VOSviewer open-access platform was chosen to explore the prominent cooperation among 219 organizations with the minimum number of documents of an organization with a minimum number of citations as five. Out of 219 organizations, 121 meet the thresholds. The largest set of connected items consists of 19 items as powerful organizations came up during the verification of the selected items almost 3%0.19 organizations were divided into three main clusters. Clusters 1 and 2 included 7 items and cluster 3 included 5 items. The highest number of publications were from "Queensland univ" following with six by "univ calif Berkeley", four by "Deakin univ", and so on.

11.3.9 Summary of Research Findings

The research on smart cities, artificial intelligence, and sustainability has seen significant growth from 2011 to 2023. Sustainability is the most productive source, with

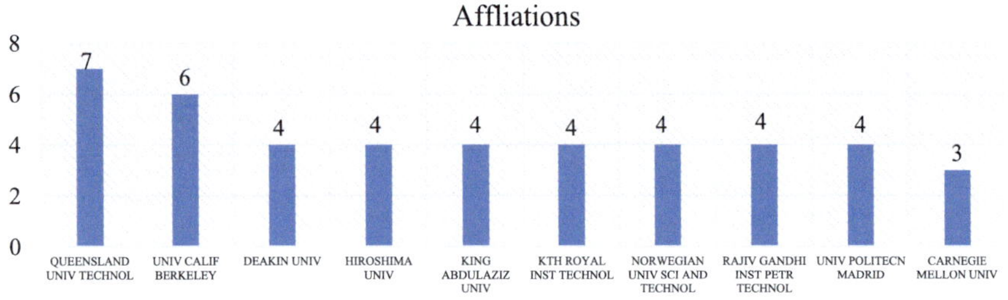

Fig. 11.8 The graph shows the network visualization of affiliations TS = (smart city and artificial intelligence and sustainability)

Yigitcanlar T as the leading author. China, the USA, and India are the top contributing countries. From 656 keywords, 72 met the network visualization threshold, with keywords like "Artificial Intelligence", "Sustainability", and "Smart Cities" showing the strongest co-occurrence. The word tree map highlights "Big Data," "Cities", and "Smart Cities" as dominant keywords, emphasizing the need for research bridging these areas to address practical habitation challenges. Out of 343 authors, 181 met the threshold for co-authorship analysis, with "Yigitcanlar T" and "Allam Z" emerging as the most influential authors. The strongest collaborations were observed among Australia, China, and Spain. Globally, "Smart City" and "Smart Cities" dominate, with limited focus on AI and sustainability. In India, "Smart Cities" and "Governance" are frequent, but gaps remain in integrating AI and IoT. Top affiliations include Queensland University of Technology, University of California Berkeley, and Deakin University.

The analysis highlights the growing intersection of the Sustainability, Artificial Intelligence (AI) and Smart Cities. Even if the advancements are quite evident, there is a significant gap persists in how well these domains are integrating to address real-world challenges. In order to bridge the gap, the study underscores the need of emphasizing the necessity of collaborative, interdisciplinary efforts for sustainable urban developments.

11.4 Future Directions of Research

Future research must address several crucial areas to fully realize the full potential of artificial intelligence (AI) in advancing sustainable smart cities. A thorough framework has to be developed to enable AI integration after the complete assessment of readiness of cities and their smooth functioning. Through multidisciplinary research, experts should focus on identifying the gaps and bridging them between sustainability, smart cities and artificial intelligence.

Investigating underrepresented subfields such as Internet of things (IoT), AI-driven governance frameworks and systems management will support a more comprehensive understanding to position their role in smart city ecosystems. Furthermore, expanding bibliometric analysis to encompass the world-wide trends and understudied topics will aid in creating globally feasible strategies for sustainable urban growth.

To address challenges and creating cutting-edge AI solutions for urban sectors like energy, e-governance, transportation and data analytics, collaboration between researchers, practitioners and legislators should be prioritized. Future studies can make a substantial impact and significantly contribute in making the cities not only technologically sophisticated but also socially and environmentally sustainable.

A thorough analysis of key factors such as infrastructure such as data centers and connectivity, citizen engagement such as public awareness, participation in smart city initiatives, policy frameworks, and technology integration. Researchers can better grasp the practical challenges and prospects in implementing artificial intelligence (AI) for sustainable smart cities development.

11.5 Conclusion

The idea of smart and sustainable cities has been introduced as a fresh, creative approach to dealing with urbanization-related problems. Adopting new technologies that make use of artificial intelligence in order to facilitate citizens with improved facilities has become essential. A number of urban sectors, including education, health, energy management, business, data analytics, transportation monitoring, security, and e-governance, are being functionally active by AI at a rapid rate. In order to conduct a thorough investigation based on bibliometric analysis of the rising drift of integrating artificial intelligence in the development of smart cities and seeking to make them sustainable, this study has been examined to look into the published documents in the last 11 years. Researchers who study smart cities are becoming more and more interested, which reflects the urgent need for sustainability and the incorporation of AI. The following significant study fields can be identified in the field of smart cities on the basis of co-keyword network analysis: (a) smart development; (b) artificial intelligence; (c) smart strategy for sustainable development; and (d) Internet of things in industry 4.0. The advantages of AI are discussed in making cities smart as well as sustainable, but there are still certain gaps that need to be resolved. The following primary critical areas should now be the focus of future efforts. First, it needs to be determined how prepared smart cities are for implementing and using artificial intelligence technologies and later the framework has to be designed to incorporate AI for successful implementation.

The study's core idea is to identify subfields like the Internet, the Internet of things, models, management, systems, and "things" have been the main focus of recent research. The exploding new keywords refer to "smart cities" and "Sustainability" and supported the idea that smart cities need to be made viable on a global scale with the use of artificial intelligence. Key terms including "artificial intelligence", "sustainability", and "IoT" were found to be less frequently together. An analysis of the most productive authors from 98 publications on smart cities, artificial intelligence, and sustainability shows that author "Yigiycanlar T" from China has published the most articles on sustainability year-by-year between 2011 and 2022. Allam Zall of the United States ranks second on IEEE access with 29 articles per year. It has been highlighted that there aren't many publications on sustainable living, artificial intelligence, or smart cities. The word tree demonstrates the disconnect between artificial intelligence and smart cities. Having a trend in smart cities, sustainability, and artificial intelligence is urgently necessary because there is still a gap to use the complete potential of AI in the holistic development of sustainable smart cities.

References

Agarwal P, Kumar RGV, Agarwal P (2020) IoT based framework for smart campus: COVID-19 readiness. In: Yang XS, Fong SJ, Toapanta M, Andronache I, Phillips N (eds) Proceedings of the 2020 fourth world conference on smart trends in systems, security and sustainability (Worlds4 2020), pp 539–542

Alahmari N, Alswedani S, Alzahrani A, Iyad K, Albeshri A, Mehmood R (2022) Musawah: a data-driven AI approach and tool to co-create healthcare services with a case study on cancer disease in Saudi Arabia. Sustainability 14(6). https://doi.org/10.3390/su14063313

Albrecht SV, Beck JC, Buckeridge David L, Botea A, Caragea C, Chi C, Damoulas T, Dilkina Beaton E, Fazli P, Ganzfried S, Giles CL, Guillet S, Holte R, Hutter F, Koch T, Leonetti M, Lindauer M, Machado MC, Malitsky Y, Zheng Y (2015) Reports on the 2015 AAAI workshop series. AI Mag 36(2):90–101. https://doi.org/10.1609/aimag.v36i2.2590

Allam Z, Dhunny ZA (2019) On big data, artificial intelligence and smart cities. Cities 89:80–91. https://doi.org/10.1016/j.cities.2019.01.032

Al-Marghilani A (2022) Artificial intelligence-enabled cyberbullying-free online social networks in smart cities. Int J Comput Intell Syst 15(1). https://doi.org/10.1007/s44196-022-00063-y

Anthony B Jr (2021) A case-based reasoning recommender system for sustainable smart city development. AI & Soc 36(1):159–183. https://doi.org/10.1007/s00146-020-00984-2

Chui KT, Alhalabi W, Pang SSH, Ordonez de Pablos P, Liu RW, Zhao M (2017) Disease diagnosis in smart healthcare: innovation, technologies and applications. Sustainability 9(12). https://doi.org/10.3390/su9122309

Chui KT, Lytras MD, Visvizi A (2018) Energy sustainability in smart cities: artificial intelligence, smart monitoring, and optimization of energy consumption. Energies 11(11). https://doi.org/10.3390/en11112869

Cugurullo F, Acheampong RA, Gueriau M, Dusparic I (2021) The transition to autonomous cars, the redesign of cities and the future of urban sustainability. Urban Geogr 42(6):833–859. https://doi.org/10.1080/02723638.2020.1746096

D'Amico G, Arbolino R, Shi L, Tan Y, Ioppolo G (2021) Digital technologies for urban metabolism efficiency: lessons from urban agenda partnership on circular economy. Sustainability 13(11). https://doi.org/10.3390/su13116043

Donthu N, Kumar S, Mukherjee D, Pandey N, Lim WM (2021) How to conduct a bibliometric analysis: an overview and guidelines. J Bus Res 133:285–296. https://doi.org/10.1016/j.jbusres.2021.04.070

Garlik B (2022) Application of artificial intelligence in the unit commitment system in the application of energy sustainability. Energies 15(9). https://doi.org/10.3390/en15092981

Öztürk O, Kocaman R, Kanbach DK (2024) How to design bibliometric research: an overview and a framework proposal. Rev Manag Sci 18:3333–3361. https://doi.org/10.1007/s11846-024-00738-0

Yigitcanlar T, Mehmood R, Corchado JM (2021) Green artificial intelligence: towards an efficient, sustainable and equitable technology for smart cities and futures. Sustainability 13(16). https://doi.org/10.3390/su13168952

MIX
Papier aus verantwortungsvollen Quellen
Paper from responsible sources
FSC® C105338

If you have any concerns about our products,
you can contact us on
ProductSafety@springernature.com

In case Publisher is established outside the EU,
the EU authorized representative is:
Springer Nature Customer Service Center GmbH
Europaplatz 3, 69115 Heidelberg, Germany

Printed by Libri Plureos GmbH
in Hamburg, Germany